LEARNING MATH THROUGH VISUAL ART AND HANDS ON PROJECTS

JAVIER S. GUERRERO
Spring Valley, California, 2004

Copyright © 2002 by Javier S. Guerrero.   26187-GUER
Library of Congress Control Number: 2004098602
ISBN 10:   Softcover   1-4134-7541-8

ISBN 13:   Softcover   978-1-4134-7541-8

All rights reserved. No part of this book may be reproduced or transmitted in any form or by any means, electronic or mechanical, including photocopying, recording, or by any information storage and retrieval system, without permission in writing from the copyright owner.

This book was printed in the United States of America.

To order additional copies of this book, contact:
Xlibris Corporation
1-888-795-4274
www.Xlibris.com
Orders@Xlibris.com

INDEX

INTRODUCTION
A. I love my school and my students
B. Different ways of learning

NUMBER SENSE

1. Writing quantities
2. Positional value of numbers and writing quantities with decimals
3. Roman numbers and project
4. Rounding numbers
5. The factorization forest
6. Divisibility rules
7. Prime factorization
8. Greatest common factor and least common denominator
9. Addition and subtraction of fractions
10. Multiplication and division of fractions
11. Working with fractions and mixed numbers
12. Mixed numbers-addition and subtraction
13. Mixed numbers-multiplication & division
14. Proportion and equivalent fractions
15. Decimals, fractions and percent
16. Percent of increase and decrease
17. Exponents and roots of 10
18. Scientific notation
19. Basic operations vocabulary
20. Properties of operations
21. Order of operations
22. Order of operations color examples
23. The sets of numbers
24. The set of integers
25. Addition of integers
26. Subtraction of integers
27. Multiplication and division of integers

STATISTICS & PROBABILITY

28. Spreadsheet database
29. Stem and leaf plot
30. Correlated graph
31. Bar graph
32. Box-and-Whisker plot
33. Circle graph
34. Line graph
35. Possibilities and Probability
36. Probability-experiment and theory
37. Creating a game based on probability

MEASUREMENT

38. Measurement using the ruler
39. Understanding the area units
40. Metric system
41. Conversion of Units chart
42. Conversion of Units machine

THE PYRAMID OF LEARNING

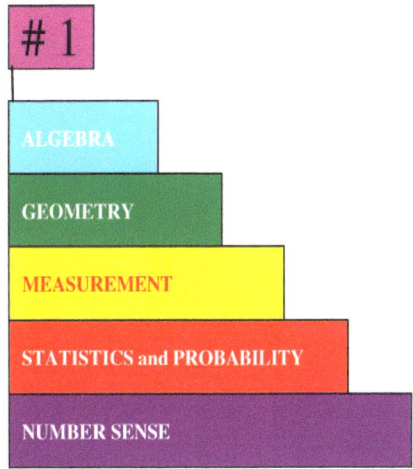

GEOMETRY

43. Regular polygons
44. Pentagon project
45. Rectangle formulas
46. Triangle formula
47. Parallelogram formula
48. Trapezoid formula
49. Classifying triangles
50. The Pythagorean Theorem
51. Circle.
52. Surface area
53. Volume of prisms and pyramids
54. Cylinder
55. Cone and cylinder

ALGEBRA

56. The Mobile
57. The language of algebra
58. Addition equations mobile
59. Multiplication equations mobile
60. Solving simple equations
61. Solving inequalities
62. Writing algebraic expressions
63. Addition in algebra
64. Multiplication in algebra
65. Division in algebra
66. Powers in algebra
67. Exponents-rules and examples
68. Two step equations with multiplication
69. Two step equations with division
70. Two step equations with like terms
71. Two step equations with like terms and fractions
72. Coordinate Plane
73. Linear Equations formulas
74. Functions
75. Discovering patterns

I  my school and my students

My students are like candles that come in many sizes, colors, and shapes but all have the capacity for light and learning. At their age, they are changing every day and their moods vary from moment to moment. They come with a variety of math backgrounds and different learning styles.

I have been a middle school math teacher for fifteen years and love it.

However, it was not love at first sight. My life as an educator started in 1969 teaching Architecture at the University level in Mexico. After immigrating to the United States and working in the business sector, I made another career change. When I began to teach at the middle school level in the United States, I was very motivated and enthusiastic. I took out the book the school gave me and began to go through the chapters related to the curriculum and felt I was doing my best to make a difference in the lives of my students. I soon discovered that many of my students arrived to class with poor math skills. They had difficulty accessing the information from the book and complained that they were bored. Their tests showed they were not learning much. After reflecting on the problem, I decided to change my teaching style and began to write the lessons that are now part of "LEARNING MATH THROUGH VISUAL ART AND HANDS-ON PROJECTS". To my surprise, the situation in my classes started to change. My students were happy to come to class. They were interested, their behavior improved and, most importantly, they were learning! The projects in this book I have developed and refined over my years in teaching. Each project reflects my training in visual arts and sculpture that I received as an Architect. I have used all of them in my class and they work! I wrote this book for my students, since I believe that knowing math is the key to their future academic success.

Javier S. Guerrero

...to my wife Carol & my family

Different ways of learning..

I hear a REGULAR PENTAGON is a polygon with five equal sides, and I understand that some students can LEARN BY LISTENING…

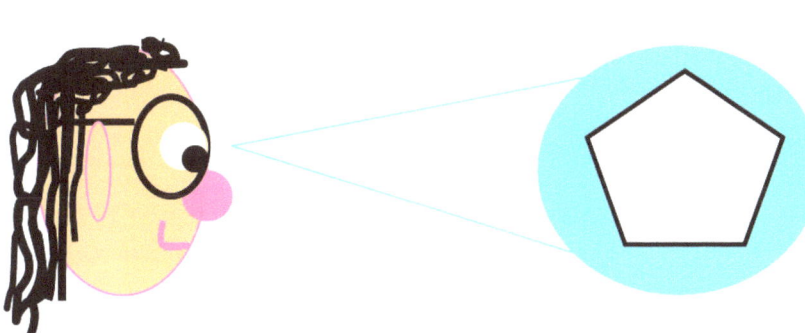

But I can only picture it if I see one. I know I am a VISUAL LEARNER.

I always remember by DOING. I like working on projects. I am a KINESTHETIC LEARNER. If we put many regular pentagons together, we create a volumetric figure.
WE LOVE DISCOVERING

…a ball?

Through the use of graphics, colors and projects, this book is designed to make learning math fun!

WRITING QUANTITIES…like writing names

Quantities can have as many as three names and one last name between commas (,) and before the decimal point (•)

NAME	QUANTITIES			
	Amount	Hundreds	Tens	Ones
John Peter Paul	1,386,214	Three hundred	eighty	six
Peter Paul	1,086,214		Eighty	six
Paul	1,006,214			Six

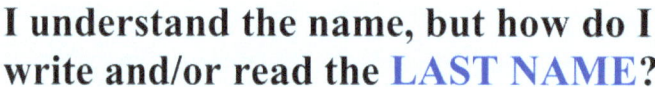

I understand the name, but how do I write and/or read the LAST NAME?

The last name depends on the position of the quantity in relation with the decimal point. See the example:

LAST NAMES:

trillions , billions , millions , thousands , ones .

3 4 6 , 2 9 7 , 8 7 5 , 6 0 4 , 0 0 9 .

346 trillions, 297 billions, 875 millions, 604 thousands, 009 ones

There is a GHOST DECIMAL POINT when there are not decimals

Writing the quantity: Three hundred forty six trillions, two hundred ninety seven billions, eight hundred seventy five millions, six hundred four thousands, nine ones

Note: If you have numbers after the decimal point, use "and" before naming the decimal number (See page on writing decimal numbers)

1

Writing Quantities with Decimal Numbers

Look at the **POSITIONAL VALUE TABLE** below, every digit to the right side of the decimal point has a special name.

1. Read and/or write the quantity of whole numbers using the name(s) and last name(s) strategy.
2. Remember to read and/or write **and** to represent **the decimal point.**
3. Read and/or write the decimal numbers just if they were another quantity of whole numbers and **finish the paragraph with the special name of the last digit.**

TRILLIONS	HUNDRED TRILLIONS	2	
	TEN TRILLIONS	5	
	TRILLIONS	1	
BILLIONS	HUNDRED BILLIONS	7	
	TEN BILLIONS	0	
	BILLIONS	4	
MILLIONS	HUNDRED MILLIONS	9	
	TEN MILLIONS	6	
	MILLIONS	3	
THOUSANDS	HUNDRED THOUSANDS	0	
	TEN THOUSANDS	0	
	THOUSANDS	8	
ONES	HUNDREDS	2	
	TENS	7	
	UNITS OR ONES	6	
DECIMAL NUMBERS	TENTHS	9	
	HUNDREDTHS	2	
	THOUSANDTHS	5	
	TEN-THOUSANDTHS	3	
	Hundred-THOUSANDTHS	1	

251 Trillions, 704 Billions, 963 Millions, 008 Thousands, 276 ones **and** 92 thousands, 531 **hundred-thousandths**

2

ROMAN NUMBERS

The roman numerals are: I = 1   X = 10   C = 100   M = 1000
V = 5   L = 50   D = 500   V̄ = 5000

Read the ROMAN numbers from left to right as we do with our numbers: **MCXI = 1111, DLV = 555, CLI = 151, MV = 1005, XI = 11**

A bar on top of any numeral represents 1000 times the number

ROMANS introduced:

The addition principle; you can repeat the numerals **I, X, C, M**, as many as three times and you add the values from left to right:
II = 2, **III** = 3, **XX** = 20, **XXX** = 30, **CC** = 200, **CCC** = 300, **MM** = 2000, **MMM** = 3000

The subtraction principle; by placing one numeral to the left of other with greater value:
IV= 4, **IX**= 9, **XL**= 40, **XLIX**= 49, **XC**= 90, **XCIX**= 99, **CDXC** = 490
CDXCIX= 499, **CM**= 900, **CMXCIX**= 999, **MCMXCIX**= 1999

NUMERALS AROUND THE WORLD are very interesting. Using color paper and twine create a project that will help you remember the numerals in other cultures. **Can you create your own numerals?**
…it is time to develop your critical thinking skills!

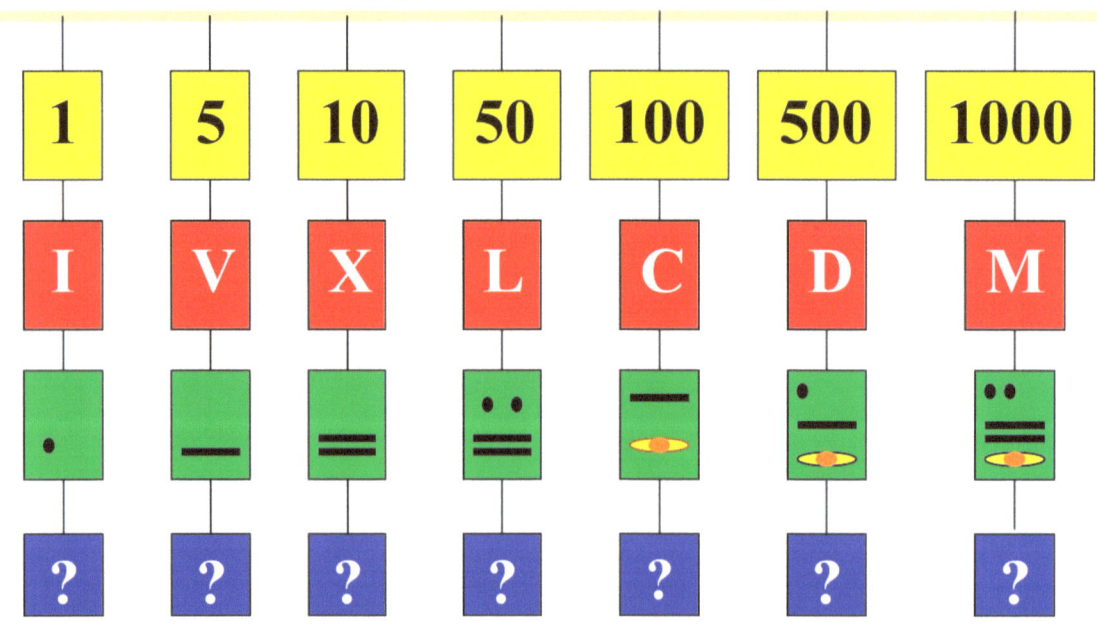

Example of THE MAYAN NUMERALS, they only used the dot, the bar and

ROUNDING NUMBERS

We are going to use the analogy of the traffic light to help you understand better rounding quantities.

Are small numbers - STOP round down

Are big numbers - GO round up

Examples of rounding down and rounding up quantities:

Round 29,451 to the nearest TEN = 29,450

When rounding down 29,451 STOP! the rounding unit doesn't change and the numbers to the right become 0

Round 8,360 to the nearest HUNDRED = 8,400

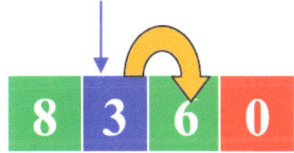

When rounding up 8,360 GO! Add 1 to the unit you are rounding   8,360
                                                                  + 1
and remember the numbers to the right become 0   8,400

Basic steps for rounding quantities:
 a) Understand the question and find the unit to be rounded using your positional value chart.
 b) Look at the number to the right side of the unit to be rounded.
 c) Round up or round down the quantity based on the big or small number to the right side of the unit that you are rounding.

THE FACTORIZATION FOREST
ONCE UPON A TIME THERE WAS A FOREST…A FACTORIZATION FOREST

In the forest live numbers…and there are two clubs of numbers:
THE PRIME NUMBERS which can only be divided by themselves and by 1
and the **COMPOSITE NUMBERS** that have more than two factors.

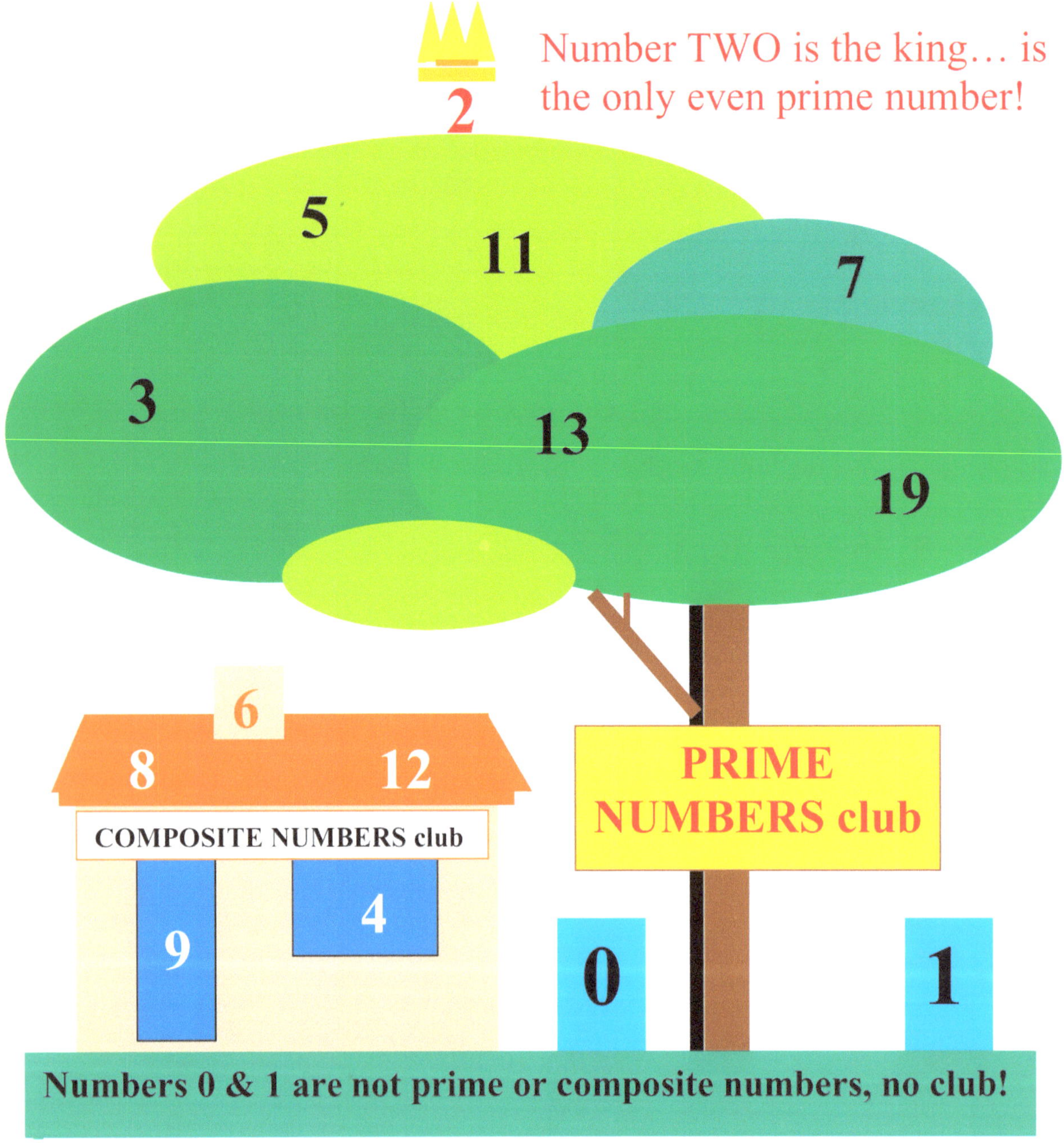

Number TWO is the king… is the only even prime number!

2

5 11 7

3 13 19

6
8 12

COMPOSITE NUMBERS club

PRIME NUMBERS club

9 4

0 1

Numbers 0 & 1 are not prime or composite numbers, no club!

DIVISIBILITY

Sorry numbers 0 and 1 are "blue", because they are neither a PRIME or COMPOSITE NUMBER

But we can still be part of the PRIME NUMBERS CLUB if we can write a number that can only be divided by itself and by one on the leaf of our tree of the "FACTORIZATION FOREST"…

We can tell whether a whole number is divisible by other whole number without actually dividing by following the rules of DIVISIBILITY

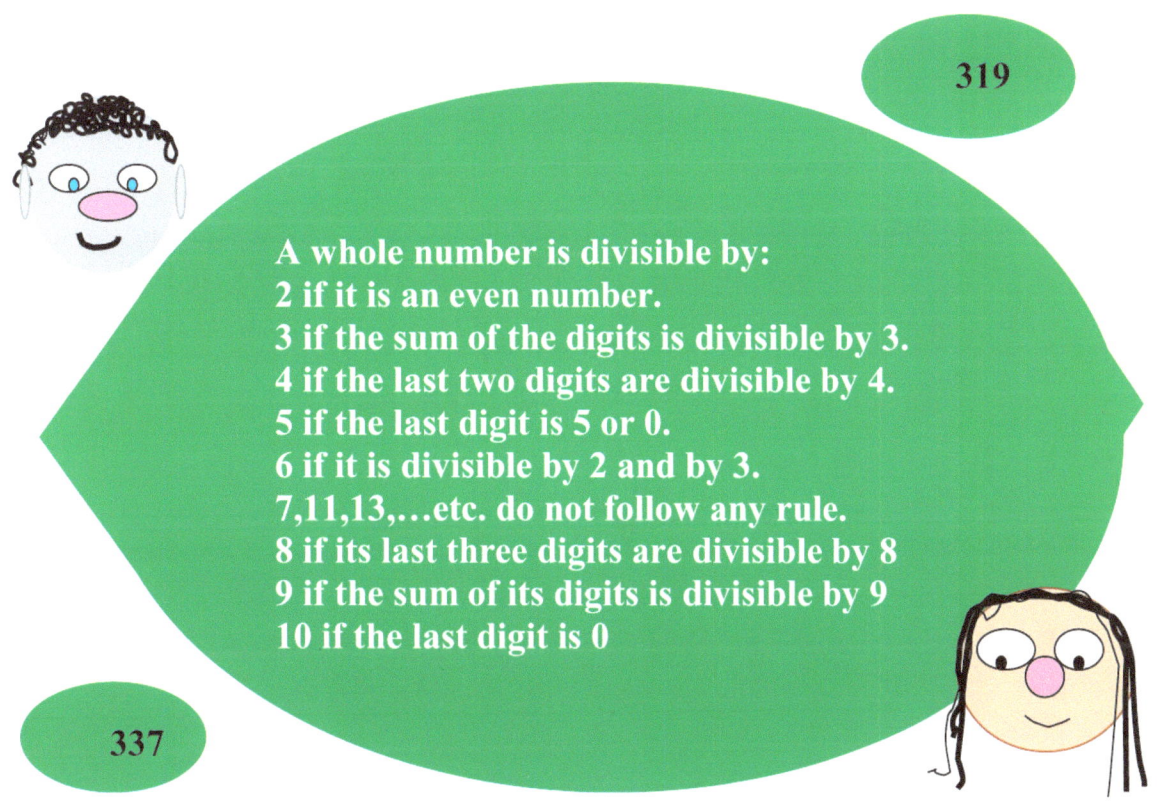

A whole number is divisible by:
2 if it is an even number.
3 if the sum of the digits is divisible by 3.
4 if the last two digits are divisible by 4.
5 if the last digit is 5 or 0.
6 if it is divisible by 2 and by 3.
7,11,13,…etc. do not follow any rule.
8 if its last three digits are divisible by 8
9 if the sum of its digits is divisible by 9
10 if the last digit is 0

Using divisibility rules to do PRIME FACTORIZATION

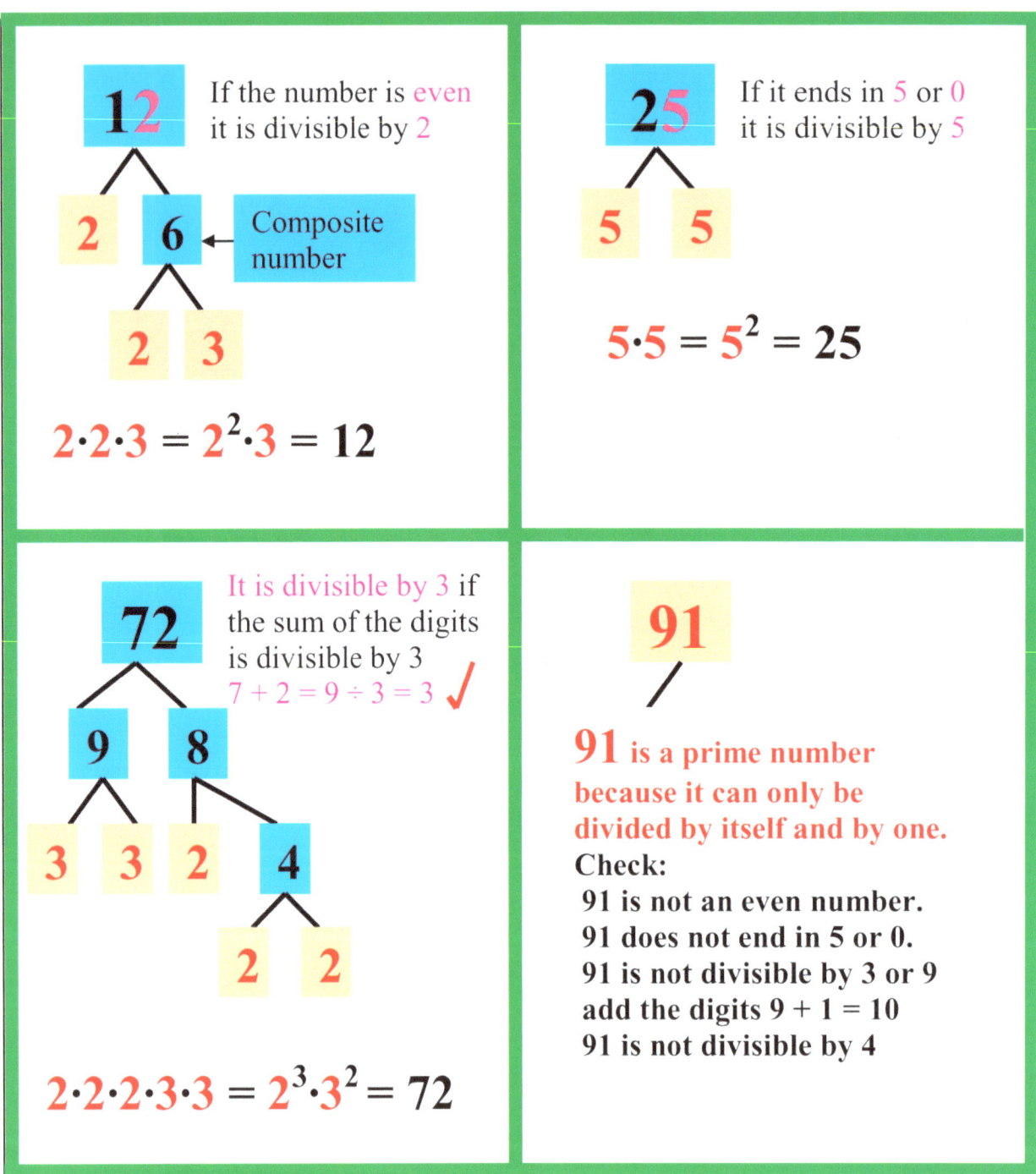

Prime Factorization is the Key for finding the Greatest Common Factor & The Least Common Denominator

GREATEST COMMON FACTOR "GCF" ("The twins")

We use it to simplify fractions…It is easy and fun!
- Do the Prime factorization of the Numerator and Denominator
- Look for the Prime Numbers "twins"; out of each pair of "Twins" only use one.
- If you have more than one pair of Prime Numbers "twins", multiply all the "Twins", but remember out of each pair of "twins" only use one number. See the examples:

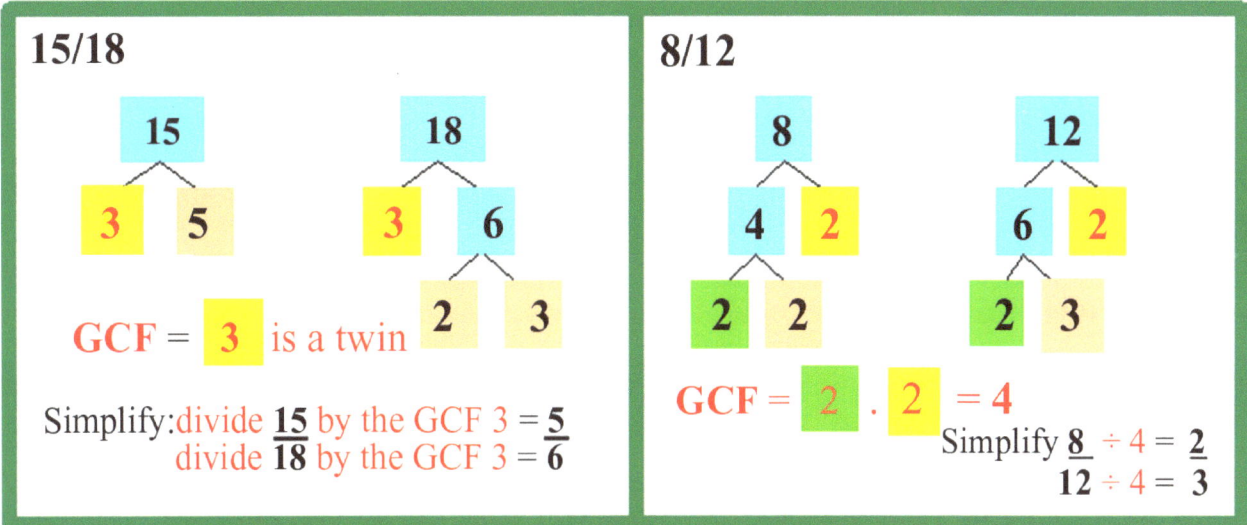

LEAST COMMON DENOMINATOR "LCD" (A party)

We use it for addition and subtraction of fractions.
(Think that the "twins" are going to have a party and they are inviting the Prime Numbers). Do the Prime factorization of the denominators.

Example: Find the "LCD" for adding **11/18 + 13/16**

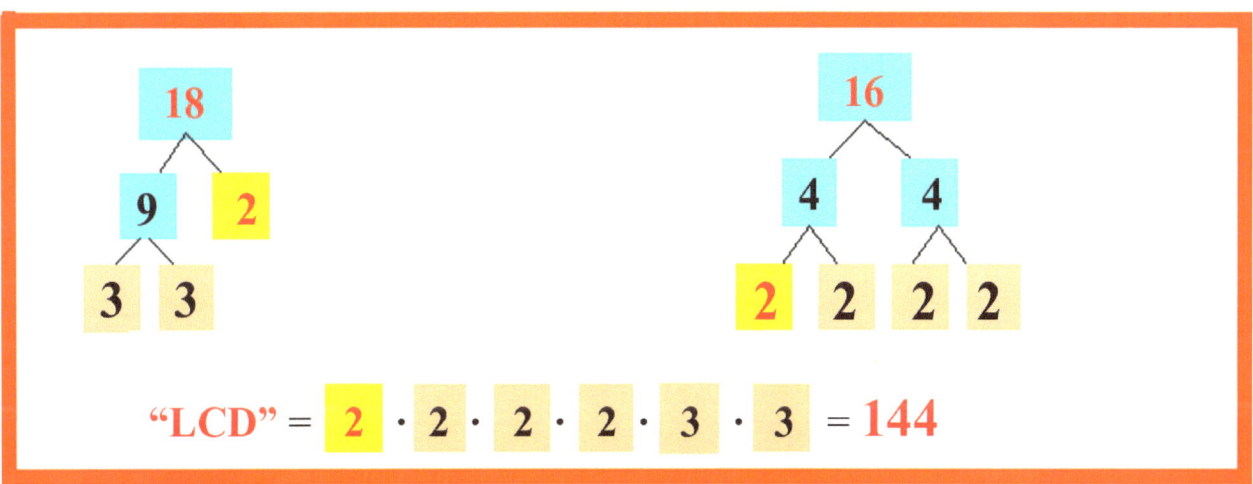

ADDITION & SUBTRACTION OF FRACTIONS

ADDITION OF FRACTIONS

$$\frac{5}{16} + \frac{7}{18} = \frac{45 + 56}{144} = \frac{101}{144}$$

Steps for addition of fractions:
- Find the **least common denominator** (see example in previous page "The Party")
- Follow the flow chart above:
  $144 \div 16 \cdot 5 = 45$ and $144 \div 18 \cdot 7 = 56$, than **add** $45 + 56 = 101$

SUBTRACTION OF FRACTIONS

$$\frac{15}{16} - \frac{3}{18} = \frac{135 - 24}{144} = \frac{111}{144}$$

Steps for subtraction of fractions:
- **Find the Least Common Denominator.**
- Follow the flow chart above:
  $144 \div 16 \cdot 15 = 135$ and $144 \div 18 \cdot 3 = 24$, than **subtract** $135 - 24 = 111$

MULTIPLICATION & DIVISION OF FRACTIONS

MULTIPLICATION OF FRACTIONS

Think of the railroad tracks

$$\frac{2}{3} \cdot \frac{3}{4} = \frac{2(3)}{3(4)} = \frac{6}{12}$$

Steps for multiplication of fractions are easy:
Multiply numerator by numerator 2(3)=6
and denominator by denominator 3(4)=12
Always simplify your answer if possible using The Greatest Common Factor "GCF":

$$\frac{6 \div \text{"GCF"} \; 6}{12 \div \text{"GCF"} \; 6} = \frac{1}{2}$$

DIVISION OF FRACTIONS

$$\frac{2}{3} \div \frac{4}{7} = \frac{2}{3} \cdot \frac{7}{4} = \frac{14}{12} = 1\frac{1}{6}$$

Steps for division of fractions:
- Flip upside down the second fraction
- Change the operation from division to multiplication and follow the steps for multiplication of fractions

Working with Fractions & Mixed Numbers

Write…

A whole number as a Fraction

$$3 = \frac{3}{1} \leftarrow \text{Ghost}$$

We use it to **multiply** Fractions with whole numbers.

Example:

$$\frac{2}{9}(4) = \frac{2}{9} \cdot \frac{4}{1} = \frac{8}{9}$$

A quantity with decimals as a Mixed Number

$$4.2 = 4\frac{2}{10}$$
tenth

$$5.01 = 5\frac{1}{100}$$
hundredths

$$8.125 = 8\frac{125}{1000}$$
thousandths

Change…

Change an Improper Fraction into a Whole or Mixed Numbers

$$\frac{12}{4} = 3 \qquad 12 > 4 \text{ makes it an Improper Fraction}$$

$$\frac{7}{3} = 2\frac{1}{3}$$

times $3\overline{)7}$
$\phantom{\text{times }3}\underline{-6}$
$\phantom{\text{times }3}\;\;1$ remainder

Change a Mixed Number into a Fraction

35 plus

$$5\frac{2}{7} = \frac{37}{7}$$

times

You will use this skill…
- Before multiplying and dividing Mixed Numbers.
- While doing subtraction of Mixed Numbers if needed.

MIXED NUMBERS – Addition and Subtraction

Addition

$2\frac{1}{4} + 5\frac{3}{8}$

Steps: Add the whole numbers.
Add the fractions.
Put them together.

$2+5=7$ and $\frac{1}{4} + \frac{3}{8} = \frac{5}{8}$ = $7\frac{5}{8}$

Some times when you add the fractions you get an improper fraction like:

2 3/4 + 5 3/8 = 7 9/8 but 9/8 = 1 1/8 so 7 + 1 1/8 = 8 1/8

and you need to change it into a mixed number for your final answer.

Subtraction

Note: Check if the mixed numbers can be subtracted as they are; if not... learn the little trick of the greater fraction!

— is greater than —...so they can be subtracted

$2\frac{5}{6} - 1\frac{2}{3}$

Steps: Subtract the whole numbers.
Subtract the fractions.
Put them together

2−1=1 and 5/6 − 2/3 = 1/6 so the answer is 1 1/6

is smaller than … so here comes the trick:

We need to change one whole into a fraction out of the first fraction, so that the first fraction is greater and you can subtract the mixed numbers following the normal steps.

$3\frac{3}{2} - 2\frac{3}{4}$ = 3−2=1 and 3/2−3/4=3/4 answer $1\frac{3}{4}$

1 ½ = 3/2 greater than

MIXED NUMBERS – Multiplication and Division

Multiplication

Steps:
- Change the Mixed Numbers into fractions.
- Multiply the fractions.
- Change your answer into a Mixed Number

$$6\frac{1}{2} \cdot \left(2\frac{3}{5}\right)$$

$$\frac{13}{2} \cdot \frac{13}{5} = \frac{169}{10} = 16\frac{9}{10}$$

Division

Steps:
- Change the Mixed Numbers into Fractions.
- Flip the second fraction upside down.
- Change the operation of division and do multiplication.

$$6\frac{4}{5} \div 2\frac{3}{8}$$

$$\frac{34}{5} \div \frac{19}{8} = \frac{34}{5} \cdot \frac{8}{19} = \frac{272}{95}$$

Simplify your improper fraction 272÷95 = 2 82/95

PROPORTION & EQUIVALENT FRACTIONS

A proportion is a statement that two ratios are equal…1/2 = 3/6.
It helps us to solve problems like:
COMPARING FRACTIONS and ORDERING RATIONAL NUMBERS
Compare 7/8 and 9/12 by following these steps:

- Find the Least Common Multiple of the denominators 8 and 12

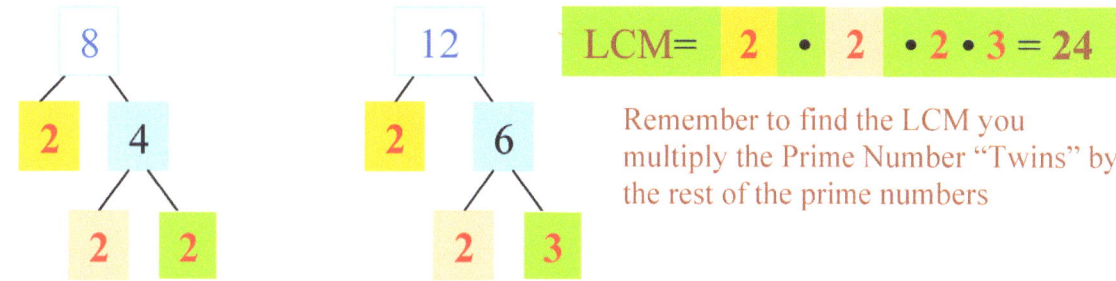

Remember to find the LCM you multiply the Prime Number "Twins" by the rest of the prime numbers

- Change 7/8 and 9/12 into fractions with the same common denominator 24

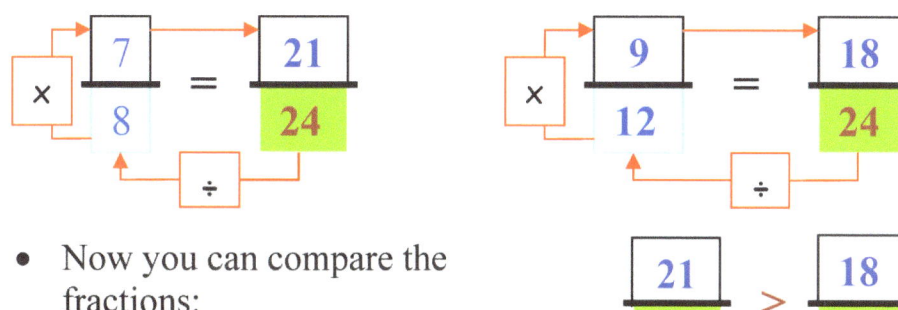

- Now you can compare the fractions:
  So 7/8 is greater than 9/12

$$\frac{21}{24} > \frac{18}{24}$$

Order in sequence from smallest to greatest and find them in the number line the following rational numbers: 7/8, ½, ¾, 5/6, ¼ and 9/15

- Check if the given line can be divided, or if it is divided in a number of parts. In our example 24. This number represents the common denominator!
- Change each fraction into that denominator 24 (some answers can be decimals):
  7/8=21/24, ½=12/24, ¾=18/24, 5/6=20/24, ¼=6/24, 9/15=14.4/24

Now you can order in sequence the fractions and find them in the number line.
Ordering from smallest to greatest: ¼, ½, 9/15, ¾, 5/6, 7/8

DECIMALS, FRACTIONS AND PERCENT

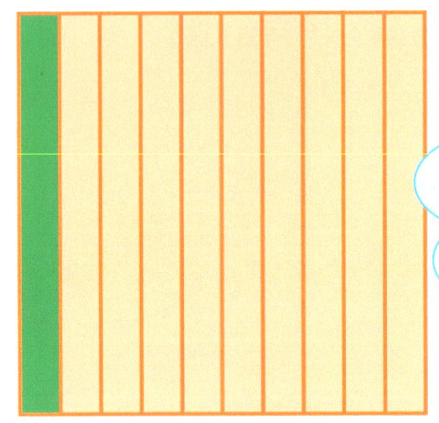

Understanding the concept of the parts and the whole is going to help me understand decimals, fractions & **percent**.

The whole is the square; it is like the price of an item at the store… The parts are 10 and 1 part of 10 is 1/10 = 0.1 = 10 percent of the whole

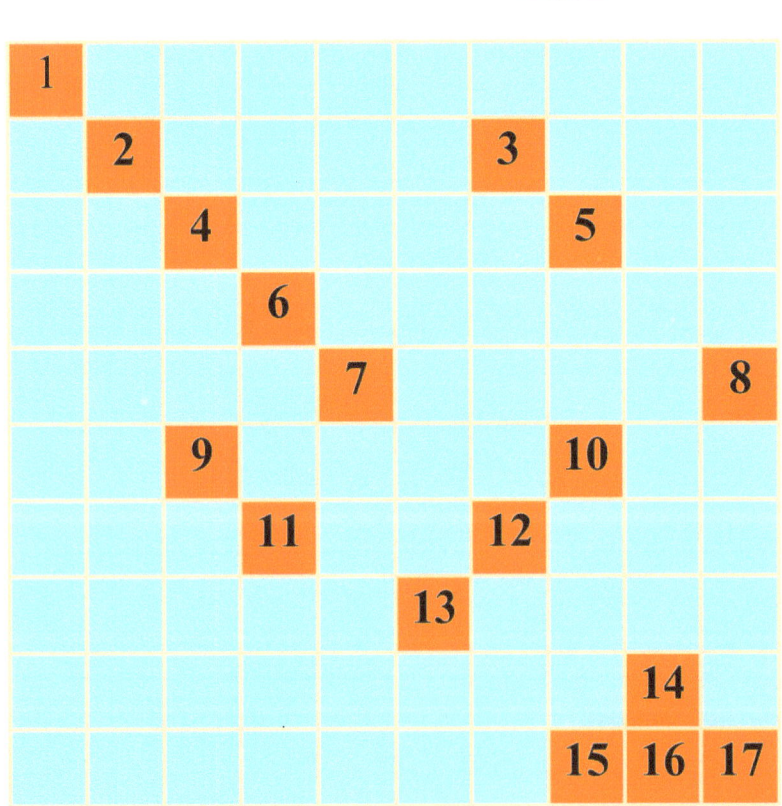

The concept of **Percent** comes from dividing **the whole in 100 parts.** It is like dividing the cost of an item at the store by 100. In the example there are 17 orange parts out of 100 17/100 = 0.17 =17 percent Imagine that the cost of a pair of shoes is $40 less 17%. How much are you paying for the shoes? Think the whole is $40 so 40÷100 = 0.40 and you are paying 100-17=83 Multiply 83(0.40)=$33.20 You are saving money… Learning about percent is going to make you smart!

PERCENT OF INCREASE AND DECREASE

> Last week I put gasoline in my car and I paid $ 1.85/gallon. Today the increase price is $ 2.18/gallon. I hate it! I wonder what is the percent of change ?

To find the percent of change follow these simple steps:
- **Find the difference between the original price and the new price:** $ 2.18 – $1.85 = 0.33
- **Divide the difference in price by the original price:** 0.33 ÷ 1.85 = 0.178
- **Round it to the hundredth:** 0.178 ∼ 0.18 = 18/100 = 18%

DISCOUNT STORE

All items on SALE

I went to the store today and I paid **$28.65** for some pants which regular price is $ 40. 99. I wonder what was my percent of decrease?

- **Find the difference between the original price and the new price:** 40.99 – 28.65 = 12.34
- **Divide the difference in price by the original price:** 12.34 ÷ 40.99 = 0.301
- **Round your answer to the hundredth:** 0.301 ∼ 0.30 = 30/100 = 30%

EXPONENTS

The **EXPONENT** or **POWER** tells us how many times the number called the **BASE** is used as a factor.

BASE 8^3 means $8(8)(8) = 512$

> Understanding powers of 10 is going to help us do Scientific Notation

$10^6 = 1,000,000$ or $10(10)(10)(10)(10)(10)$
$10^5 = 100,000$ or $10(10)(10)(10)(10)$
$10^4 = 10,000$ or $10(10)(10)(10)$
$10^3 = 1,000$ or $10(10)(10)$
$10^2 = 100$ or $10(10)$
$10^1 = 10$
$10^0 = 1$
$10^{-1} = 0.1$ or $1/10$ one ten**th**
$10^{-2} = 0.01$ or $1/10(1/10) = 1/100$ one hundred**th**
$10^{-3} = 0.001$ or $1/10(1/10)(1/10) = 1/1,000$
$10^{-4} = 0.0001$ or $1/10(1/10)(1/10)(1/10) = 1/10,000$
$10^{-5} = 0.00001$ or $1/10(1/10)(1/10)(1/10)(1/10)$
$10^{-6} = 0.000001$ or $1/1,000,000$

ROOTS

The root of a number is that number which, when multiplied by itself a given number of times, will equal the given number.
The square root has a ghost 2, any other root will have a #

You can write a root as a fraction exponent

ghost

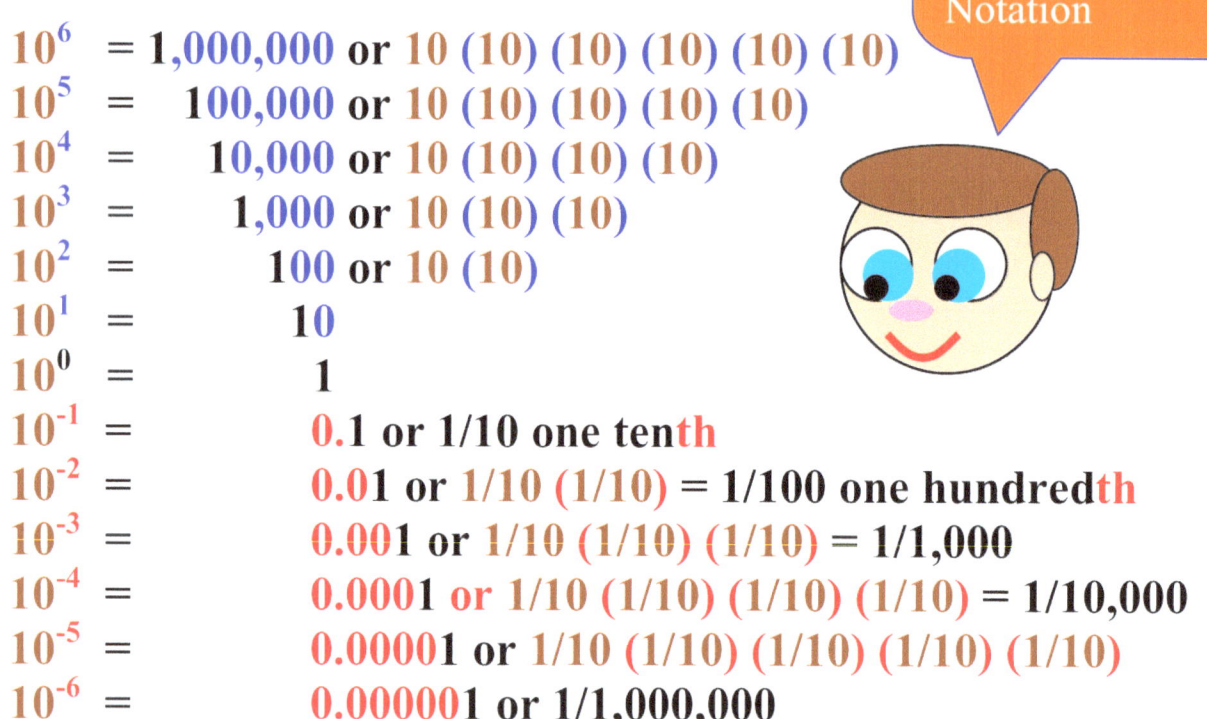

$\sqrt[2]{100} = 100^{1/2} = 10$ because $10(10) = 100$

$\sqrt[3]{1,000} = 100^{1/3} = 10$ because $10(10)(10) = 1000$

SCIENTIFIC NOTATION

Scientific notation helps us write very large or very small quantities. You write the numbers in scientific notation as the product of two factors by using powers of 10.

Example: Write the large quantity 3,800,000,000 in scientific notation.

3,800,000,000. Make the first significant digit THE UNIT ghost decimal point

3.800,000,000. Write the other digits as DECIMALS.
← + Nine

$3.8(10^9)$ Use your powers of 10 by writing a positive exponent equal to the number of places you move the decimal point

$3.8 \cdot 10^9 = 3.8 (1,000,000,000) = 3,800,000,000$

Example: Write the small quantity 0.000000086 in scientific notation.

0.000000086 Make the first significant digit THE UNIT.

0.00000008.6 Write the other digits as DECIMALS.
→ - Eight

$8.6(10^{-8})$ Use your powers of 10 by writing a negative exponent equal to the number of places you move the decimal point

$8.6 \cdot 10^{-8} = 8.6 (0.00000001) = 0.000000086$

Check your answers using your calculator!

BASIC OPERATIONS VOCABULARY

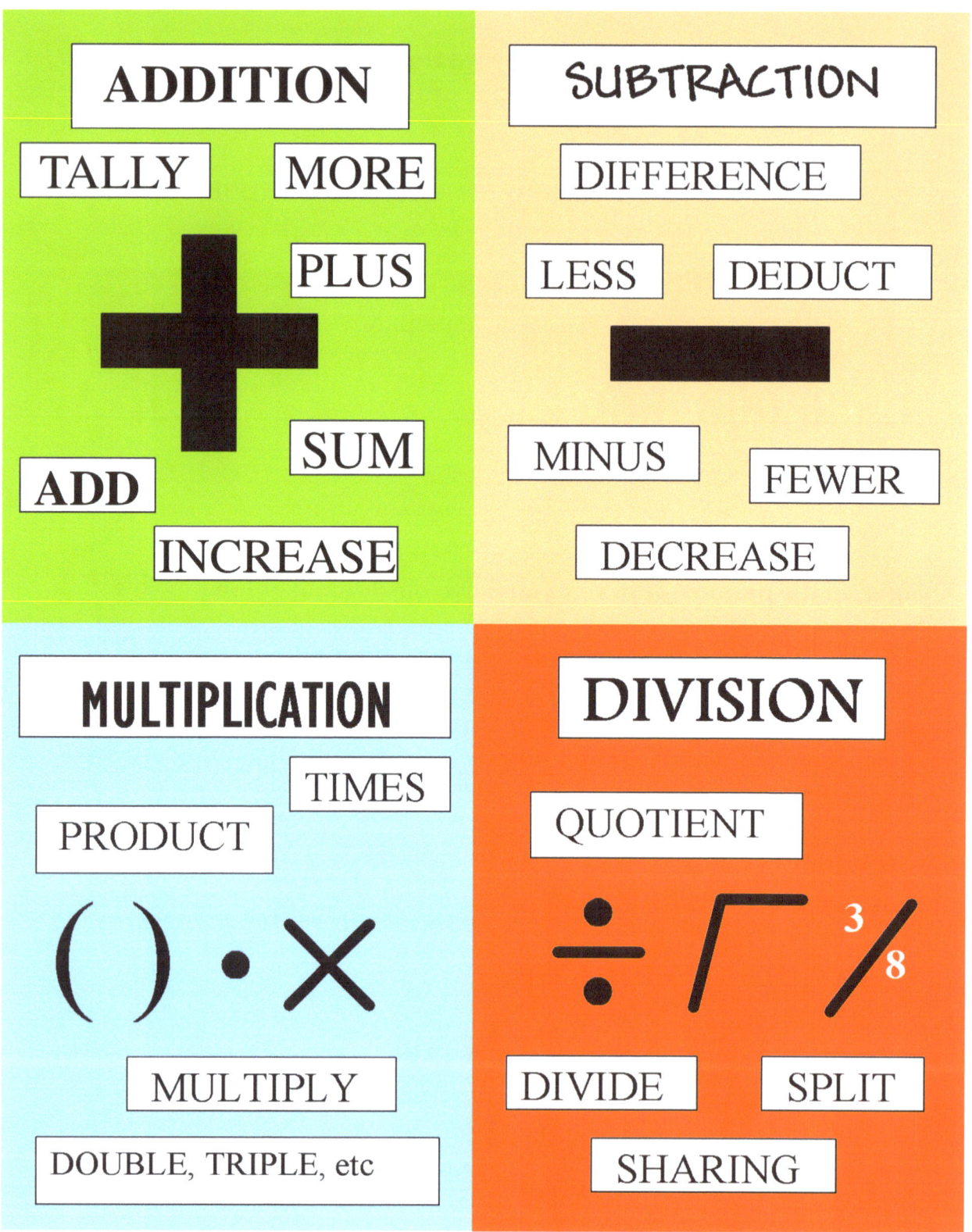

PROPERTIES OF OPERATIONS

COMMUTATIVE property of **ADDITION:** In a sum, you can add terms in any order...the answer is the same:
Examples: $3 + 4 = 7$ or $4 + 3 = 7$, $a + b = b + a$
COMMUTATIVE property of **MULTIPLICATION:** In a product, you can multiply the factors in any order...the answer is the same.
Examples: $5(8) = 40$ or $8(5) = 40$, $xy = yx$

ASSOCIATIVE property of **ADDITION:** Changing the grouping of terms does not change the sum...the answer is the same. Examples: a)
$7+(8+3) = 18$ is the same as $(7+8)+3 = 18$
$x + (y + z) = (x + y) + z$
ASSOCIATIVE property of **MULTIPLICATION:** Changing the grouping of factors does not change the product...the answer is the same.
Examples: a) $(ab)c$ is the same as $(ac)b$
b) $5(7 \cdot 2) = 70$ is the same as $(5 \cdot 2)7 = 70$

DISTRIBUTIVE PROPERTY: It combines the operations of multiplication and addition. It will help us write equivalent expressions for solving math problems...
Examples:
a) $5(3)+5(7)+5(15) = 5(3+7+15)$ ...It is easier to find the answer if we add the terms before multiplying!
b) $x(y+z) = xy + xz$ c) $2(x + 5) = 2x + 2(5) = 2x + 10$
d) $a(2a + 5) = 2a^2 + 5a$

IDENTITY property of **ADDITION:** The sum of a number or a variable an ZERO the answer is the number or the variable
Examples: a) $5 + 0 = 5$ b) $a + 0 = a$
IDENTITY property of **MULTIPLICATION:** The product of a number or a variable and ONE is the number or the variable
Examples: a) $9(1) = 9$ b) $a \cdot 1 = a$

ORDER OF OPERATIONS

1. GROUPING SYMBOLS…Evaluate expressions inside
If you have 2 or more symbols do your work **from inside ▶ out:**
Parentheses first, Brackets second, last Keys

$$\{ \leftarrow [\leftarrow (\leftrightarrow) \rightarrow] \rightarrow \}$$

THE FRACTION BAR is also a grouping sign:

$$\frac{6+2}{5-1} = \frac{8}{4} = 2$$ Do the numerator first, after do denominator and **last do the division**

2. EXPONENTS or ROOTS
The **Exponent** tell us how many times we multiply the **BASE**

$$4^3$$ Means multiply the base 4 three times
$4 \times 4 \times 4 = 64$

3. MULTIPLY or DIVIDE from left to right

$\times \rightarrow \div$ or $\div \rightarrow \times$

$12 \div 4 \times 3 =$ (division is to the left ∴ $12 \div 4$) $= 3 \times 3 = 9$
$8 \times 6 \div 12 =$ (multiplication is to the left ∴ 8×6) $= 48 \div 12 = 4$

4. ADD or SUBTRACT from left to right

$+ \rightarrow -$ or $- \rightarrow +$

$18 - 4 + 9 =$ (subtraction is to the left ∴ $18 - 4$) $= 12 + 9 = 21$
$8 + 15 - 6 =$ (addition is to the left ∴ $8 + 15$) $= 23 - 6 = 17$

ORDER OF OPERATIONS EXAMPLES

It is very important to go step by step while doing operations. The color code is going to help you follow the sequence for doing these problems:

$5^2 - 14 \div 7 =$
$25 - 14 \div 7 =$ Do the exponent first $5^2 = 5 \cdot 5 = 25$
$25 - 14 \div 7 =$ Do the division $14 \div 7 = 2$
$25 - 2 = 23$ Do the subtraction

$(24 - 18) \cdot 8 \div 2 =$
$(24 - 18) \cdot 8 \div 2 =$ Do parentheses first $24 - 18 = 6$
$6 \cdot 8 \div 2 =$ Do the multiplication $6 \cdot 8 = 48$
$48 \div 2 = 24$ Do the division

$7^3 - 7 \cdot 3 \div 7$
$343 - 7 \cdot 3 \div 7 =$ Do the exponent first $7^3 = 7 \cdot 7 \cdot 7 = 343$
$343 - 7 \cdot 3 \div 7 =$ Do the multiplication $7 \cdot 3 = 21$
$343 - 21 \div 7 =$ Do the division $21 \div 7 = 3$
$343 - 3 = 340$ Do the subtraction

$5 [14 - (8 + 4) \div 6] =$ Do all the operations inside the bracket
$5 [14 - (8 + 4) \div 6] =$ Do the parentheses first $8 + 4 = 12$
$5 [14 - 12 \div 6] =$ Do the division $12 \div 6 = 2$
$5 [14 - 2] =$ Do the subtraction $14 - 2 = 12$
$5 [12] = 60$ Do the multiplication

…remember a number outside the bracket means multiplication.

SETS { a collection of objects or numbers } inside the key sign.

- Sets are named with capital letters: A, B, C, D, E, F, G,...X, Y, Z
- Subsets are part of a SET and are also represented with capital letters: A, B, C, D, F, G, H, I, J, K, L,...X, Y, Z

⊂ means Subset and ⊄ means not a Subset

All the numbers inside the SET are Elements of the Set: A = { 5,6,7,8 }

∈ means Element of the set and ∉ means not an Element of the Set

{ } and ∅ means Empty Set, Null Set (no solution)

In Mathematics we are going to deal with several Sets of numbers that we can represent in the Line of Numbers:

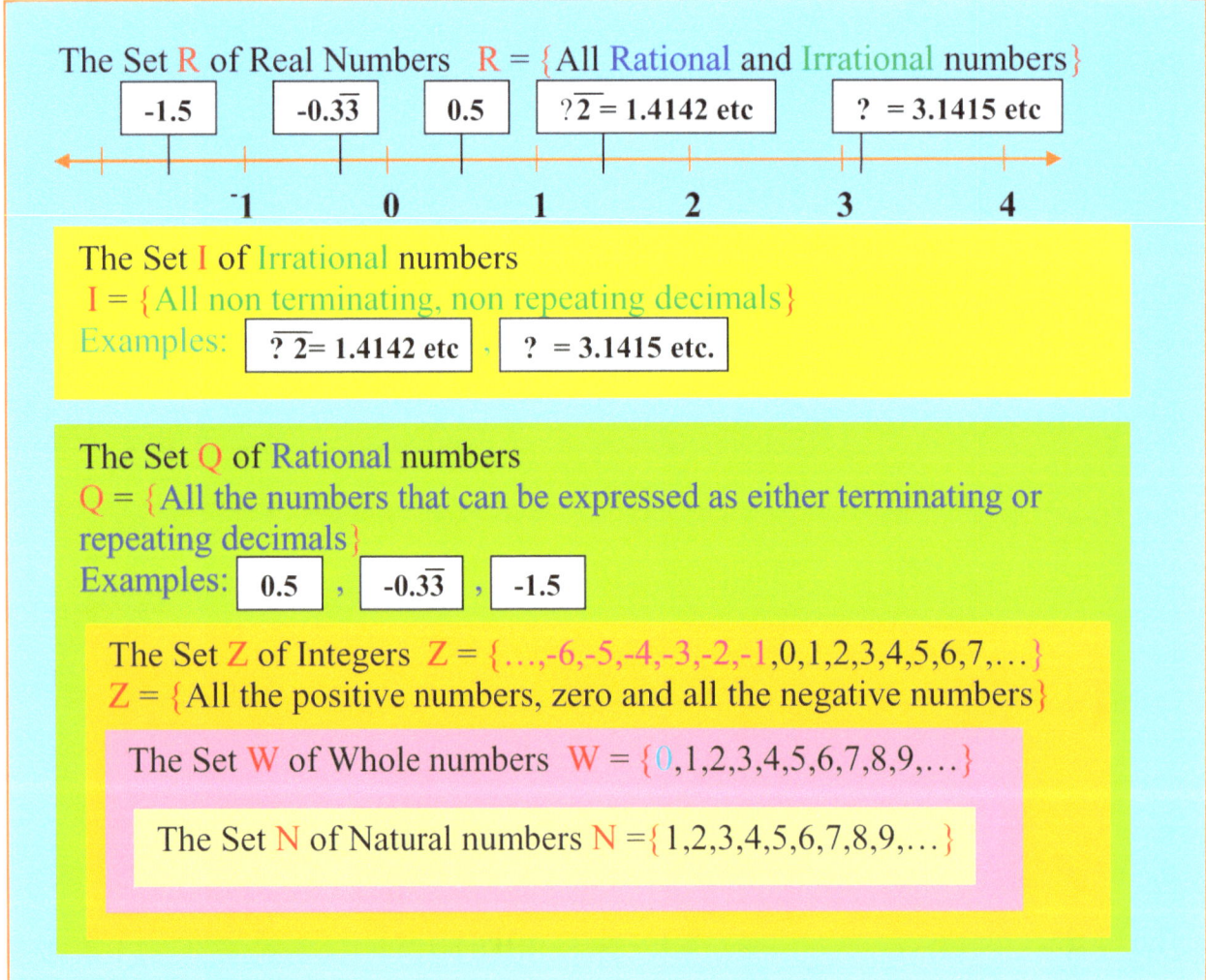

The Set R of Real Numbers   R = {All Rational and Irrational numbers}

-1.5, -0.33̄, 0.5, √2 = 1.4142 etc, π = 3.1415 etc

The Set I of Irrational numbers
I = {All non terminating, non repeating decimals}
Examples: √2 = 1.4142 etc , π = 3.1415 etc.

The Set Q of Rational numbers
Q = {All the numbers that can be expressed as either terminating or repeating decimals}
Examples: 0.5 , -0.33̄ , -1.5

The Set Z of Integers   Z = {...,-6,-5,-4,-3,-2,-1,0,1,2,3,4,5,6,7,...}
Z = {All the positive numbers, zero and all the negative numbers}

The Set W of Whole numbers   W = {0,1,2,3,4,5,6,7,8,9,...}

The Set N of Natural numbers   N = {1,2,3,4,5,6,7,8,9,...}

THE SET OF INTEGERS

Integers are the whole numbers and their opposites (negative)

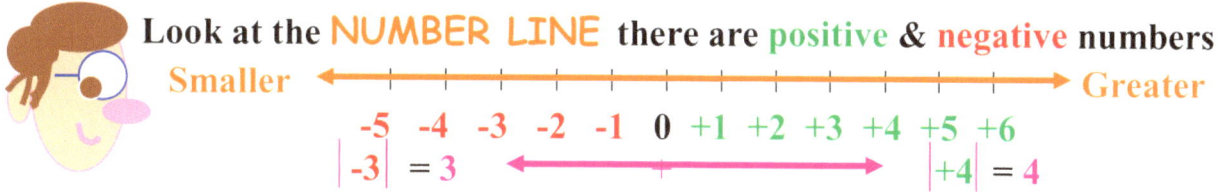

A number's distance from zero is the absolute value.

Writing Integers we use a little positive or negative sign. The number ZERO does not have a sign. Any number other than ZERO that does not have a sign is positive.

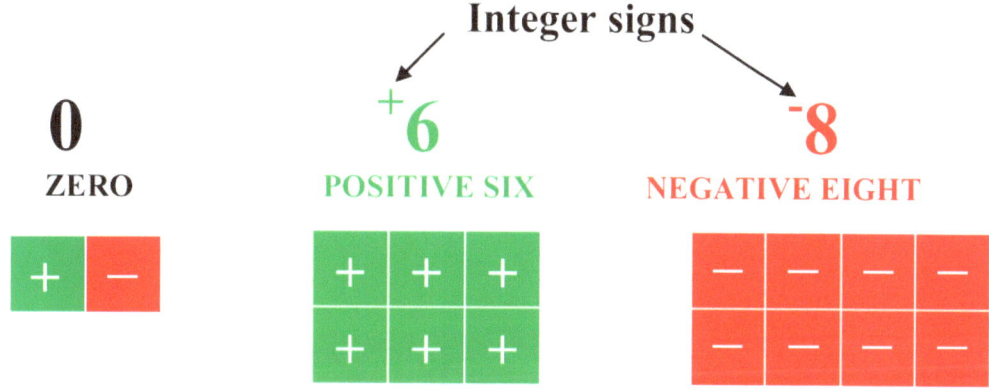

COMPARING INTEGERS.

Looking at the NUMBER LINE:
- The numbers get BIGGER going to the RIGHT
- And SMALLER going to the LEFT

$^+6$ is greater than $^+5$ It is farther to the right side in the NUMBER LINE

$^+6 > {}^+5$ the sign > means GREATER THAN

$^-4$ is smaller than $^-2$ It is farther to the left side in the NUMBER LINE

$^-4 < {}^-2$ the sign < means SMALLER THAN

0 is greater than $^-3$ It is farther to the right side in the NUMBER LINE

$0 > {}^-3$ 0 is a very important number!

ADDITION OF INTEGERS... Where is the money?

I have $2 in my left pocket and $3 in my purse.

$^+2 + {}^+3 = {}^+5$ I have a total of **$5**

Rule: The sum of two positive integers is always positive

I borrowed $1 from my sister today and $3 yesterday.

$^-1 + {}^-3 = {}^-4$ I borrowed a total of **$4**

Rule: The sum of two negative integers is always negative

I have $5 in my wallet but I am paying $1 for my soda

$^+5 + {}^-1 = {}^+4$ I still have **$4**

I have only $2 and I need to pay $4 for my book

$^+2 + {}^-4 = {}^-2$ I need **$2** more

Steps for adding positive ➕ and negative ➖ integers:

- Ignore the signs of the integers and subtract the small number from the large number.
- Give the sign of the large number to the answer

SUBTRACTION OF INTEGERS ...Take away

The concept of Zero is very important for subtraction of integers, we can write Zero with different amount of units. **Use your positive and negative cards. See the examples:**

[+][−] = 0, [+][+][−][−] = 0, [+][+][+][−][−][−] = 0

Basic steps for doing subtraction of integers:
- **Change the operation of subtraction to addition**
- **Change the sign of the subtracted integer**
- **Do the addition of integers**

$$^-4 - {^-2} = {^-2} \implies {^-4} + {^+2} = {^-2}$$

[−][−] [−][−] Take away 2 negative = $^-2$

$$^+3 - {^+1} = {^+2} \implies {^+3} + {^-1} = {^+2}$$

[+][+][+] Take away 1 positive = $^+2$

$$^-2 - {^+1} = {^-3} \implies {^-2} + {^-1} = {^-3}$$

[+][−] [−][−] Create a 0, take away [+] = $^-3$

$$^+3 - {^-2} = {^+5} \implies {^+3} + {^+2} = {^+5}$$

[+][+] [+][+][+] [−][−] Create a 0, take away [−][−] = $^+5$

MULTIPLICATION and DIVISION of INTEGERS

The rules for multiplication and division of integers are the same:
- **Same** signs equal **positive** answer.
- **Different** signs equal **negative** answer.

Multiplication... What about the money?

My parents are going to double my $3 allowance.

$^+2 \cdot {}^+3 = {}^+6$

Now I am going to get $6

I need to pay for 3 tickets for the dance at school at $2 each

I am paying a total of $6

$^+3({}^-2) = {}^-6$

Different signs = negative
$^-4({}^+8) = {}^-32$, $\;{}^+7 \cdot {}^-9 = {}^-63$

Same signs = positive
$^-3 \cdot {}^-5 = {}^+15$, $\;{}^+6({}^+4) = {}^+24$

Division

My cousin and I are sharing

We are getting $3 each

$^+6 \div {}^+2 = {}^+3$

Each of us is paying $4

$^-8 \div {}^+2 = {}^-4$

We are splitting an $8 bill

Different signs = negative
$^-8 \div {}^+2 = {}^-4$, $\;{}^+12 \div {}^-3 = {}^-4$

Same signs = positive
$^-36 \div {}^-6 = {}^+6$, $\;{}^+9 \div {}^+3 = {}^+3$

SPREADSHEET DATABASE FOR 30 COUNTRIES

Name of the Country	Population in millions	AREA in thousands of square miles	Gross Domestic Product "GDP" in Billions	Gross Domestic Product Per Capita	Life Expectancy Males	Life Expectancy Females	Literacy % of the Population
Afghanistan	26	250	20	800	48	47	32
Argentina	37	1,068	374	10,300	71	79	96
Australia	19	2,968	394	22,700	77	83	99
Brazil	173	3,286	1,040	6,100	59	68	85
China	1,261	3,705	4,420	3,600	69	72	82
Costa Rica	4	1,068	374	10,300	71	79	95
Cuba	11	42	17	1,560	73	78	96
Egypt	68	387	188	2,858	61	65	51
El Salvador	6	8	18	3,000	67	74	71
England	60	95	1,252	21,200	75	80	100
Ethiopia	64	435	33	560	39	42	35
France	59	211	1,320	22,600	75	83	99
Germany	83	138	1,810	22,100	74	81	100
Ghana	20	92	34	1,800	55	60	64
India	1,104	1,269	1,689	1,720	63	65	52
Iraq	23	169	52	2,400	66	68	58
Israel	6	8	102	18,100	77	81	96
Italy	58	116	1,180	20,800	76	82	97
Japan	127	146	2,900	23,100	77	83	100
Kenya	30	225	44	1,550	46	47	78
Kuwait	2	7	44	22,700	75	80	79
Mexico	100	761	815	8,300	69	76	90
Nigeria	123	356	106	960	52	54	57
Pakistan	142	310	270	2,000	59	61	38
Philippines	81	116	271	3,500	64	70	95
Russia	146	6,952	593	4,000	59	72	99
Spain	40	195	646	16,500	74	82	97
U.S.A.	249	3,096	9,937	31,500	74	79	100
Somalia	7	246	4	600	45	48	24
Switzerland	7	16	192	26,400	76	82	100

STEM- and- LEAF PLOT.

Gross Domestic Product per capita of 30 countries of the world

STEM is the greatest place value common to all data.
LEAVES are the next place value of data.

STEM in thousands	LEAVES
0	560, 600, 800, 960
1	550. 560, 720, 800
2	000, 400, 850
3	000, 500, 600
4	000
5	median
6	100
7	
8	300
9	
10	300, 300
11	
12	
13	
14	
15	
16	500
17	
18	100
19	
20	800
21	200
22	100, 600, 700, 700
23	100
24	
25	
26	
27	
28	
29	
30	
31	500

Statistics is part of math:
- Deals with collection of information called "**Data**"
- And the way we represent it, by doing different drawings called **plots** or **graphs**

The important information is called the **measure of central tendency**:
- **Median** is the number in the middle after you organize the data from least to greatest.
- **Mode** is the most popular data. It can be one, or more.
- **Mean** is the sum of the data divided by the number of items in the data.
- **Range** is the difference between the greatest value and the least value of the data.

Measure of central tendency in the Stem-and Leaf plot:
Median: is between **4000 & 6100** = $\frac{4000+6100}{2}$ = **5,050**

Mode: **10,300 & 22,700**
Mean: $\frac{313,600}{30 \text{ countries (number of items in the data)}}$ = **10,453.33**

Range: 31,500 – 560 = **30,940**

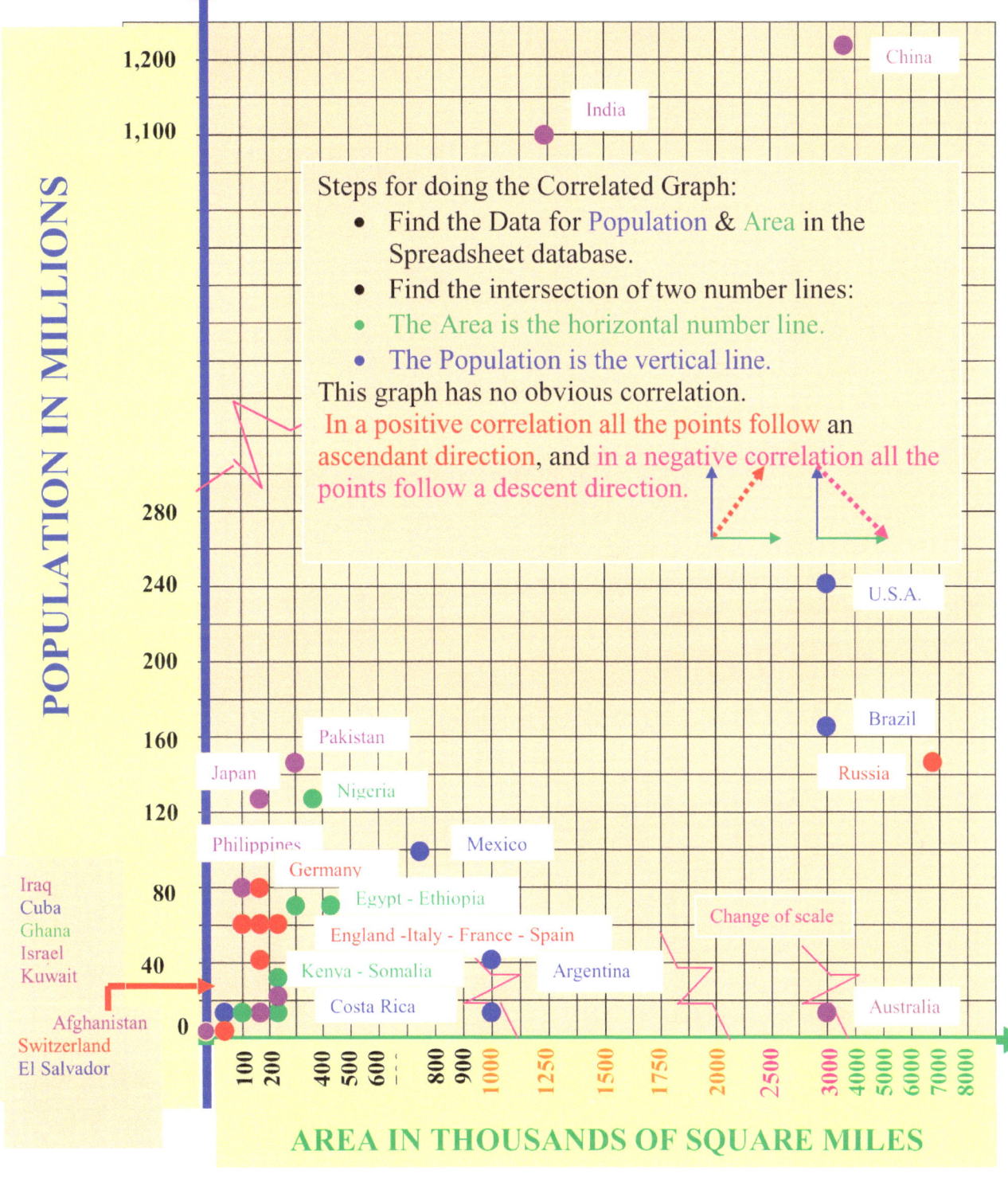

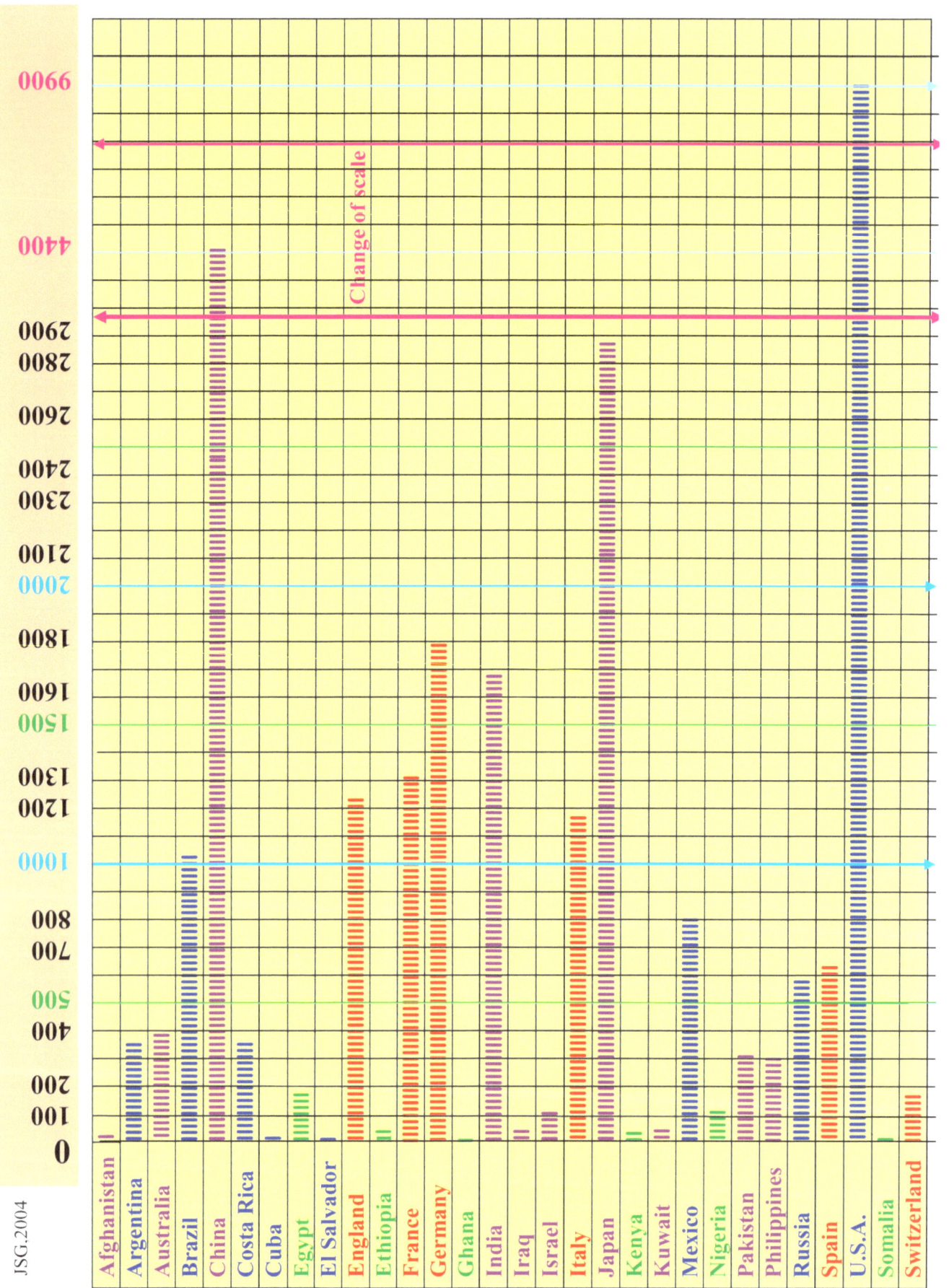

BOX-and-WHISKER PLOT
Population of 30 world countries.

Lower extreme **Lower quartile** **Median** **Upper quartile** **upper extreme**

Ok 2,4,6,6,7,7,11 **19** 20,23,26,30,37,40, **58 59** 60,64,68,81,83,100 **123** 127 142 146 173 249 **1104, 1261**
1,261

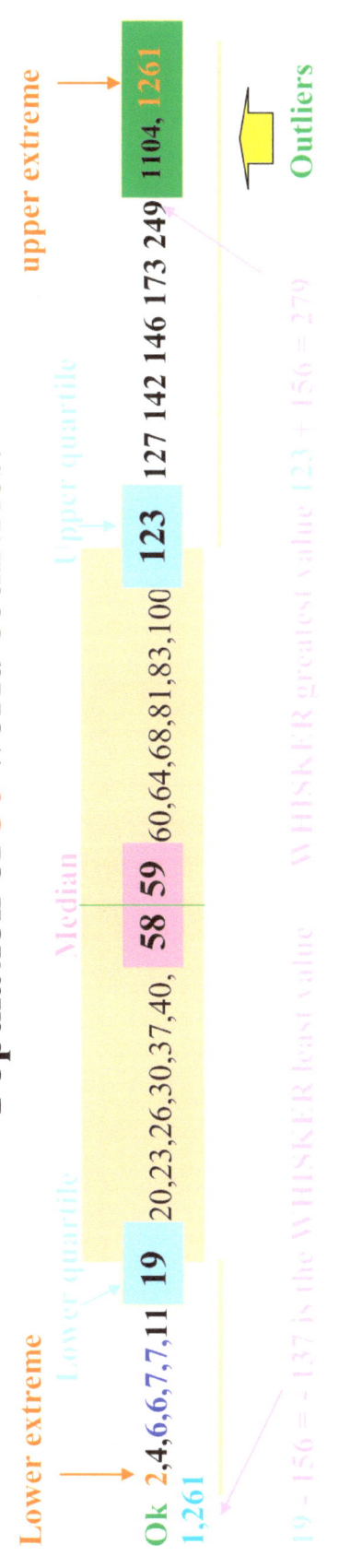

Outliers

19 – 156 = – 137 is the WHISKER least value   WHISKER greatest value 123 + 156 = 279

INTER-QUARTILE RANGE is the range of the middle half of a set of numbers: $123 - 19 = 104$

WHISKERS help us determine the least value and the greatest value of the data, and help us find the outliers of our data.

WHISKER = Inter-quartile range multiplied by a 1.5 factor.

WHISKER = $104 \,(1.5) = 156$

OUTLIERS: are the unique or special information of our data. In our example it shows the contrast of the population of China and India with the rest of the world.

RANGE: $1261 - 2 = 1259$

MODE: 6 & 7

MEDIAN: $\dfrac{58 + 59}{2} = 58$

MEAN: $\dfrac{4136}{30} = 137.87$

30 countries of theworld

Steps for doing the Box-and-Whisker Plot:
- Organize your Data from least to greatest.
- Find the Median (divide the Data in half)
  - Find the Lower Quartile (find the middle of the lower half)
  - Find the Upper Quartile (find the middle of the upper half)

You have your Box!   You can find the Inter-quartile range and the Whisker

CIRCLE GRAPH

Literacy in 30 countries of the world (See the Spreadsheet Database)

Steps for doing a Circle Graph:
- We are going to represent 30 countries in a CIRCLE Graph.
- Since the CIRCLE has 360 degrees, each country is 360/30=12 degrees.
- Multiply your number of countries by 12 degrees for each group.

Groups of different levels of literacy:

Less than 40 % literacy	4 countries (12) = 48 degrees
Between 50 and 59 % literacy	4 countries (12) = 48 degrees
Between 60 and 69 % literacy	1 country (12) = 12 degrees
Between 70 and 79 % literacy	3 countries (12) = 36 degrees
Between 80 and 89 % literacy	2 countries (12) = 24 degrees
Between 90 and 99 % literacy	11 countries (12) =132 degrees
100 % literacy	5 countries (12) = 60 degrees

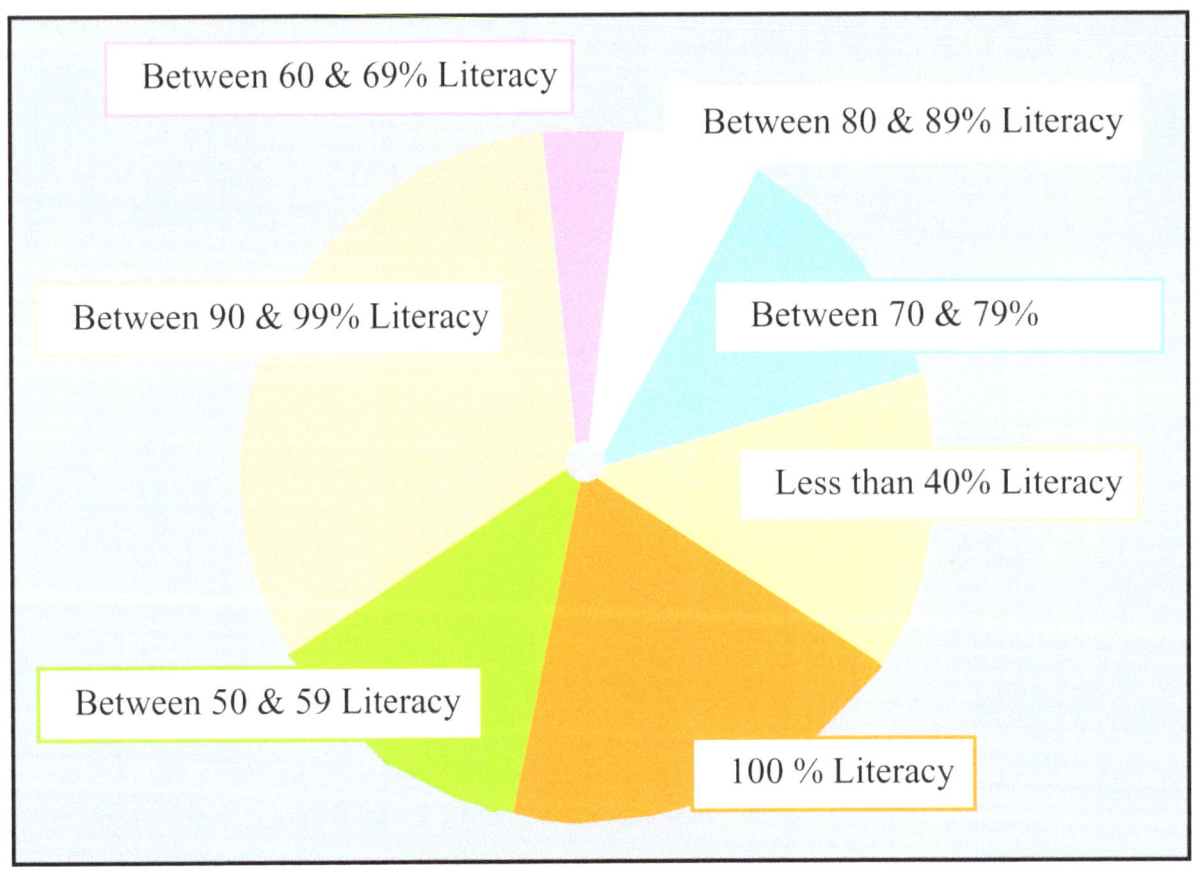

LINE GRAPH

Life expectancy for MALES " ● " and FEMALES " ○ " of 30 Countries of the World.

RANGE: from 39 to 83 = 44 (Subtract the smallest from the greatest number of the data: 83-39=44)

MEDIAN: 60+61 divided by 2 = **60.5** (The number in the middle; if there are two, add and divide by 2)

MODE: **74** (The most popular; if there is more than one as popular, all of them are mode)

MEAN: **39**+42+45+46+47+47+48+48+52+54+55+59+59+59+**60**+**61**+61+63+64+65+65+66+67+68+68+69+69+70+71+72+73+**74**+**74**+**74**+75+75+76+76+77+77+78+79+79+80+80+81+82+82+83+**83** divided by 60 ≈ **68.12** (The numerical average)

Draw a line and represent the data using a different color for males and females.
Start with the smallest number from the data and end your line with the greatest number from the data.

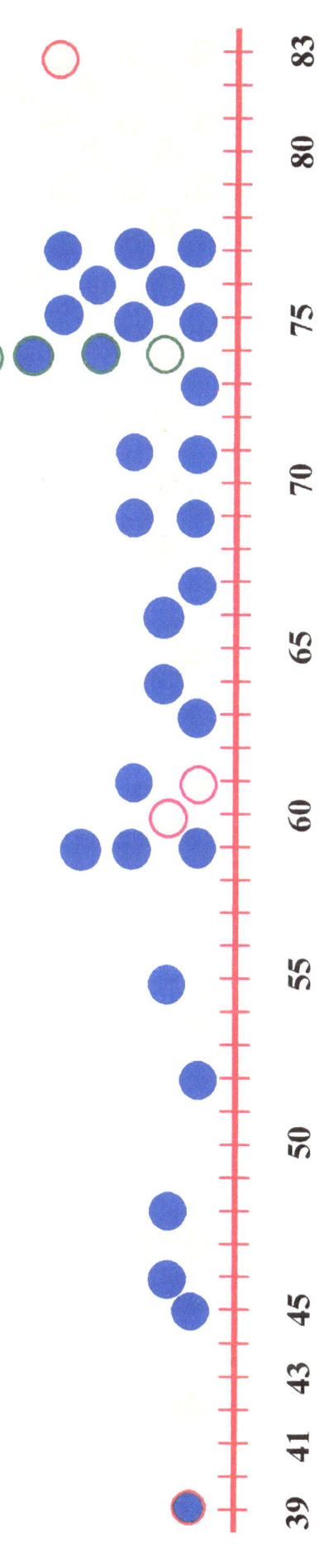

PROBABILITY

Probability is the number of times something will probably occur over the range of possible occurrences known as the possibilities, expressed as a ratio.

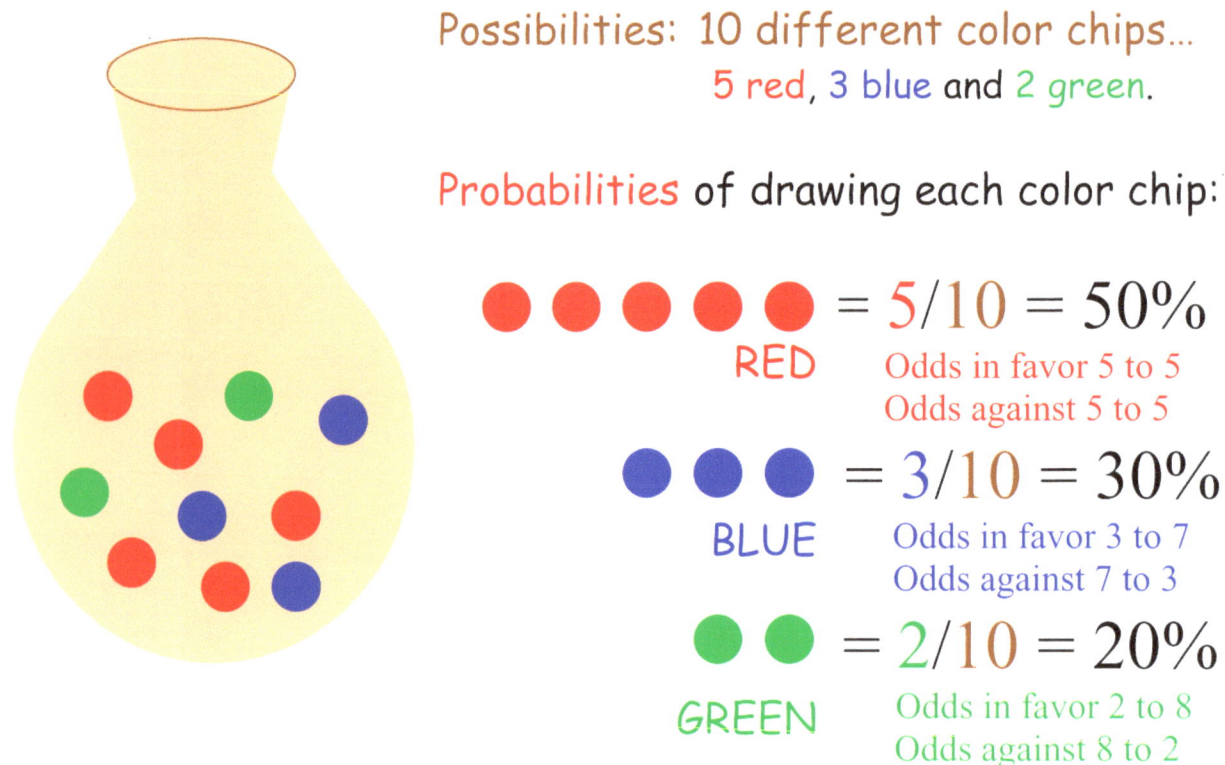

Possibilities: 10 different color chips...
5 red, 3 blue and 2 green.

Probabilities of drawing each color chip:

●●●●● = 5/10 = 50%
RED
Odds in favor 5 to 5
Odds against 5 to 5

●●● = 3/10 = 30%
BLUE
Odds in favor 3 to 7
Odds against 7 to 3

●● = 2/10 = 20%
GREEN
Odds in favor 2 to 8
Odds against 8 to 2

Comparing the THEORY with the EXPERIMENT...
From the bag of chips, draw out one chip without looking. Tally the color and RETURN the chip to the bag. Shake the bag and draw out another chip. Record the color and return the chip to the bag. Repeat the process of drawing, tallying, and returning the chip a total of 100 times. Write your results as PROBABILITIES in %

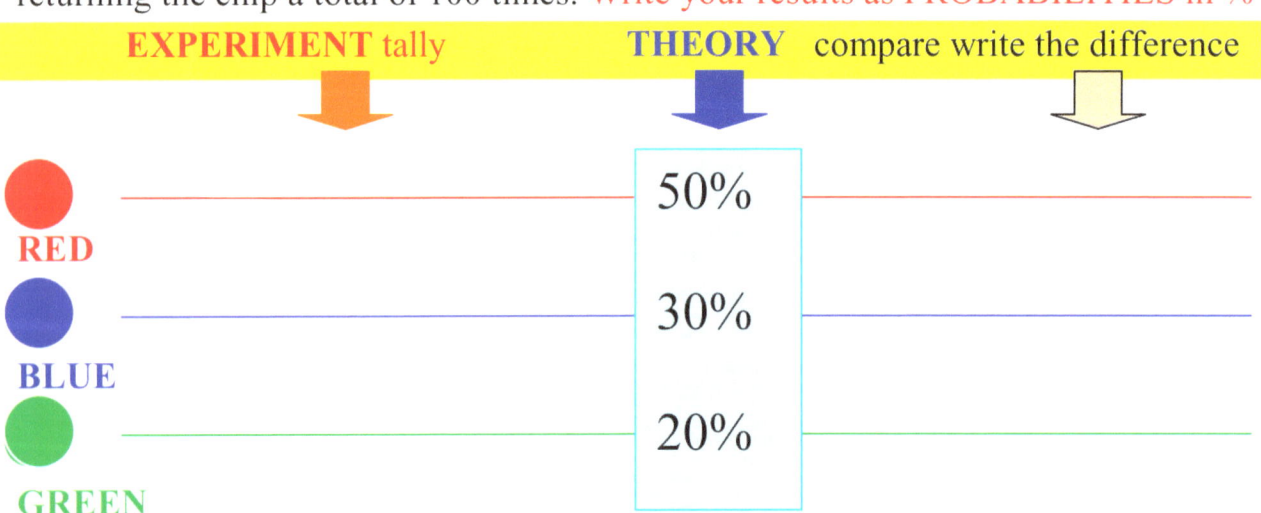

Odds in favor means: your chances of winning. You compare your bet with others.
Odds against means: your chances of loosing. You compare others bet with yours.

POSSIBILITIES AND PROBABILITIES
DICE GAME –Experiment & theory

Steps for doing the experiment of probability with two dice:
- Roll two dice at the same time and record the total of their addition each time. Examples: the least is 1+1=2 and the greatest 6+6=12
- Repeat the experiment 100 times
- Tally the results and find the percent by dividing each tally by 100. Example: ЖЖ // = 7÷100 = 7%
- **Compare the experiment with the theory**

NUMBERS	NUMBER OF POSSIBILITIES FOR ROLLING EACH NUMBER from the least 1+1=2 till the greatest 6+6=12 THE TOTAL AMOUNT OF POSSIBILITIES for rolling all the numbers from 2 through 12 with two dice is 36 (See table below) The percent of the theory can be obtain by dividing the possibilities of each number ÷ 36										THEORY	Experiment
2	1	1	Example: number 2 can only be rolled one way, so 1÷36=0.277≈ 3%								3%	
3	1	2	2	1						2÷36=0.055≈ 6%	6%	
4	1	3	3	1	2	2				3÷36=0.083≈ 8%	8%	
5	1	4	4	1	2	3	3	2			11%	
6	1	5	5	1	2	4	4	2	3	3	14%	
7	1	6	6	1	2	5	5	2	3	4	4	3	17%	
8	2	6	6	2	3	5	5	3	4	4	14%	
9	3	6	6	3	4	5	5	4			11%	
10	4	6	6	4	5	5					8%	
11	5	6	6	5							6%	
12	6	6									3%	

CREATING A GAME BASED ON PROBABILITY

In baseball there are some plays that happen more often than others.
Based on your understanding of probabilities, create a chart that reflects the probability of each play starting with 1 for the plays that are the most unlikely to happen and working your way up for the plays more likely to occur.

Add all your numbers to find the total possibilities of your game.

Create a spinner by dividing a circle in as many parts as the total amount of possibilities (remember that the circle has 360 degrees) …so 360 degrees divided by the total amount of possibilities will give you the degrees for each possibility. Create the arrow for the Spinner and you are ready to play the game. Write down what are the probabilities for each play on your game.

Example: We have 24 possibilities in total, we divide the circle 360 by 24 = 15°

O **Home run** 2/24 ~ 8%

O **Triple** 1/24 ~ 4%

O **Grand slam** 1/24 ~ 4%

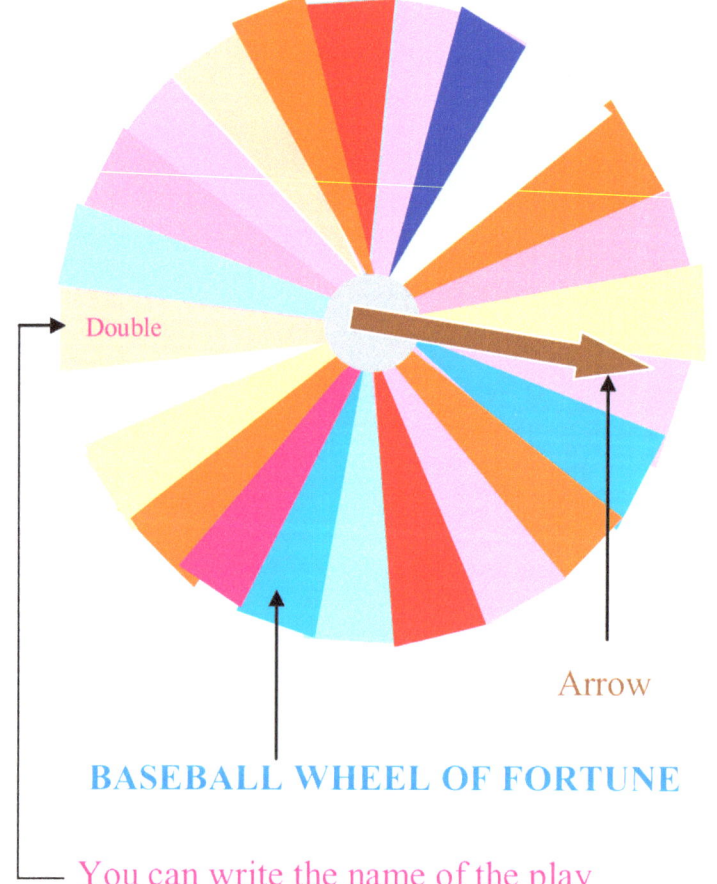

Double 3/24 ~ 13%

Hit 4/24 ~ 17%

O Steal 2/24 ~ 8%

Base on balls 2/24 ~ 8%

O Out 4/24 ~ 17%

O Double play 2/24 ~ 8%

Triple play 1/24 ~ 4%

Strike out 2/24 ~ 8%

BASEBALL WHEEL OF FORTUNE

You can write the name of the play

O O O O O O The colors represent the different plays

MEASUREMENT - Using the ruler project

Steps for doing the project:

- Using your ruler measure in **INCHES** (") the following SQUARES:
  7 1/2", 5 5/8 ", 4", 2 1/4", 5/16" and cut them in the same color.
- Measure in **CENTIMETERS** (cm) the following SQUARES :
  16.5 cm , 12 cm , 7.8 cm , 3.6 cm and cut them out in the same color.
- Make sure that the color of the centimeters squares is different than the color of the inches squares.

Example not to scale.

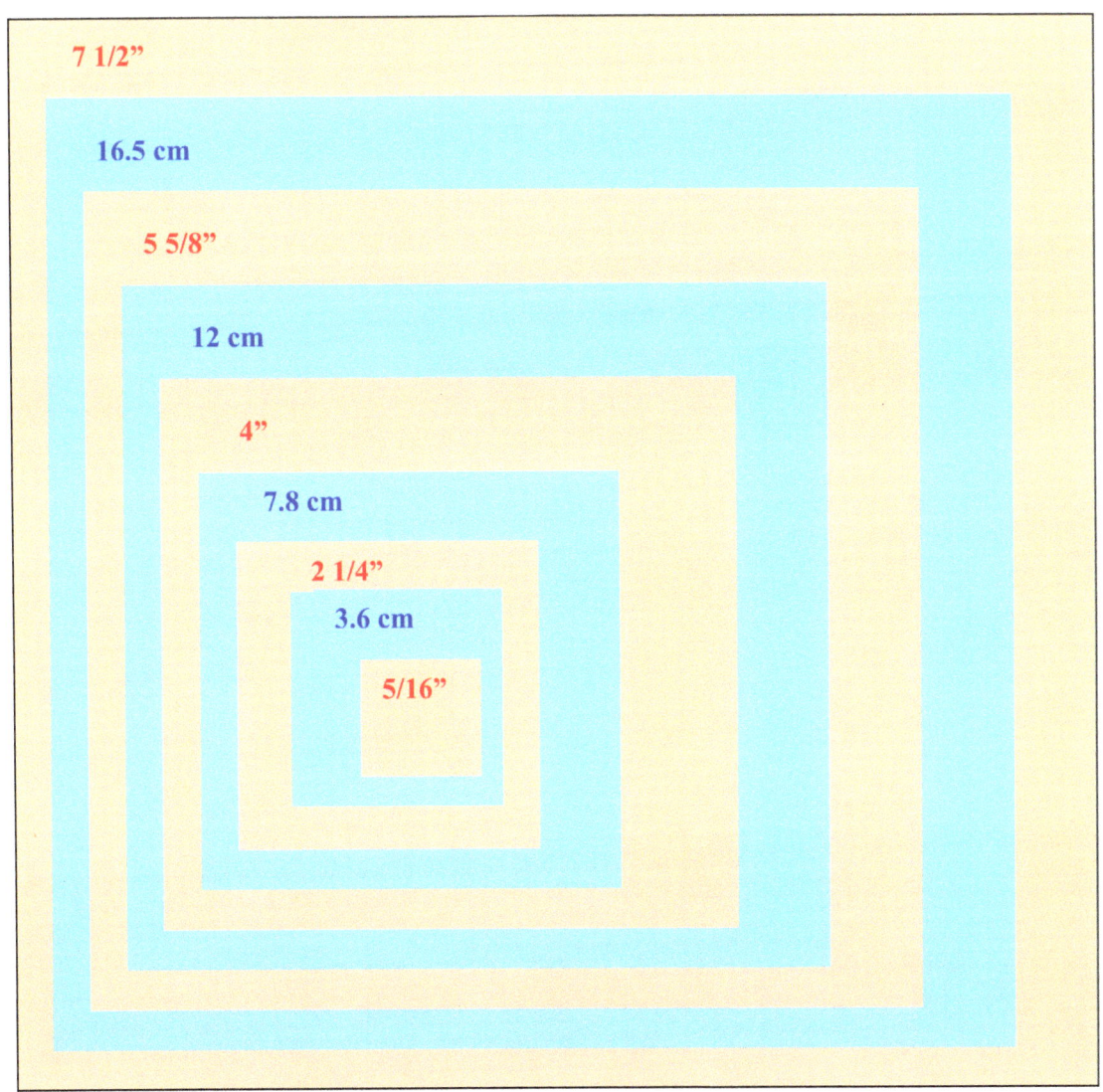

Understanding the AREA UNITS

Cut in different color construction paper using your ruler:
9 square feet to make 1 square yard, remember each square feet is 12 by 12 inches.
Paste inside four square foot: 1 square decimeter (10 by 10 centimeters), 1 square inch and 1 square centimeter.
Add in a different color of paper that area needed to make 1 square meter.

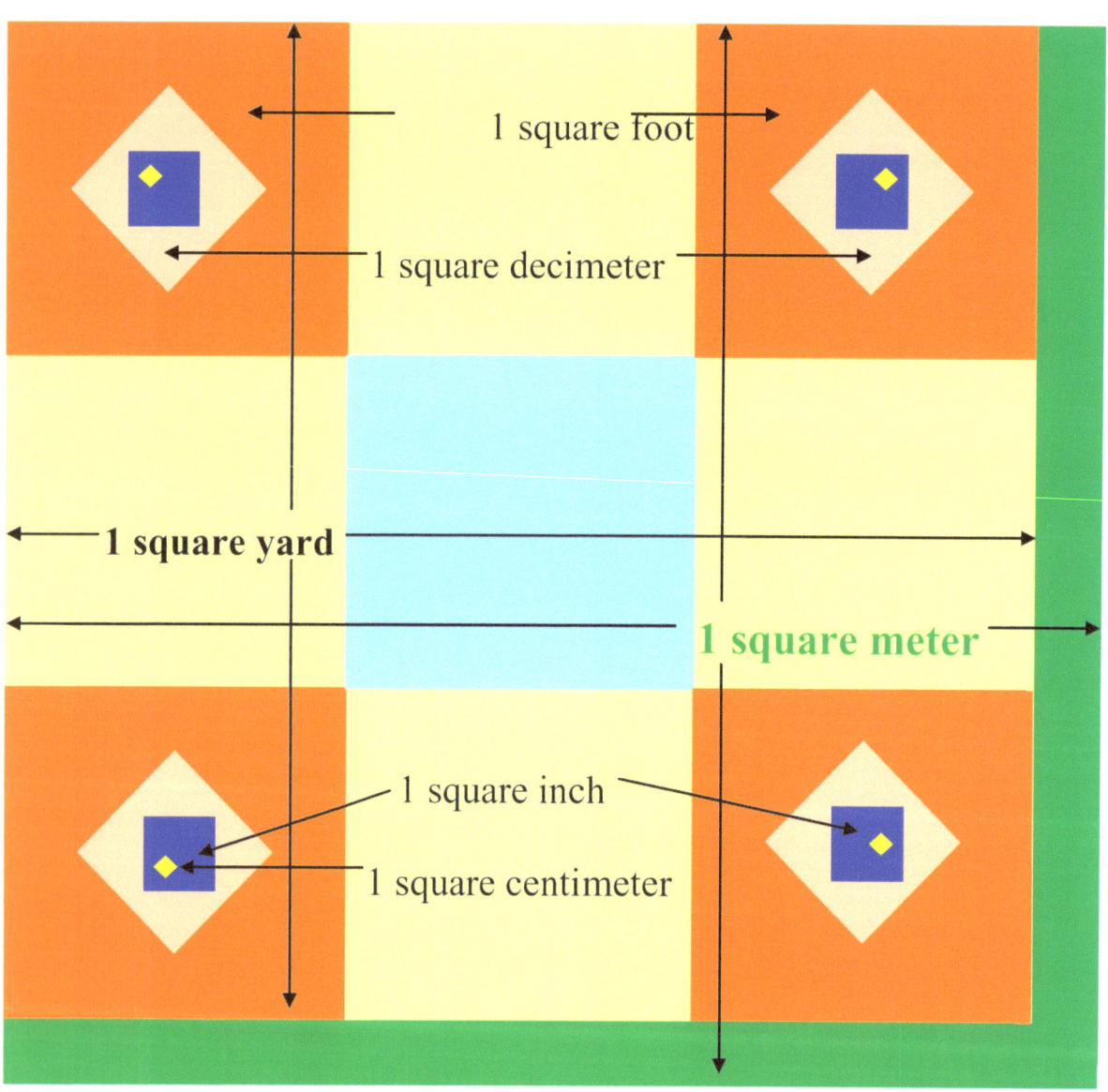

This is a great group project, you can work as a team of 9 students; each will make a square foot including the other units, after you put it together you will have you square yard. Add the extra area to make it a square meter. Can you figure out the area of your classroom in the different units?

Using a chart to solve METRIC SYSTEM problems:

Instructions for using the chart:
- Write the number that represents the KNOWN UNIT in the correct column. Example: in 879 Km. The number 9 belongs in the Km column.
- Write ZERO on the column of THE QUESTION UNIT.
- If you start your quantity with ZERO remember to write the DECIMAL POINT after ZERO.
- Complete the chart by filling up with zeroes the spaces between the known unit and the question unit

LENGTH – *LONGITUD*						m = meter	
Problem: Write…	Km	Hm	Dam	m	dm	cm	mm
2 Km in dm	2	0	0	0	0		
4 m in mm				4	0	0	0
12 Hm in dm	1	2	0	0	0		
214 Dam in mm	2	1	4	0	0	0	0
9 cm in Hm		0.	0	0	0	9	
18 dm in Km	0.	0	0	1	8		
318 mm in Dam			0.	0	3	1	8

VOLUME – *CAPACIDAD*						l = liter	
Problem: Write…	Kl	Hl	Dal	l	dl	cl	ml
7 Hl in ml		7	0	0	0	0	0
12 Kl in dl	12	0	0	0	0		
436 Hl in l	43	6	0	0			
9 Liters in kl	0.	0	0	9			
12 cl in Hl		0.	0	0	1	2	
314 ml in Dal			0.	0	3	1	4
7 dl in l				0.	7		

WEIGHT – *PESO*						g = gram		
	T			Kg	Hg	Dag	g	dg	cg	mg
5 T in Kg	5	0	0	0					
3 Kg in g				3	0	0	0		
2 Hg in cg					2	0	0	0	0

CONVERSIONS OF UNITS CHART
for English and Metric Systems

ENGLISH SYSTEM - CUSTOMARY UNITS
1 foot (ft) = 12 inches (in), same as 1' = 12"
1 mile (mi) = 1,760 yards (yd) = 5,280 feet (ft)
1 table spoon (T) = 3 teaspoons (t)
1 pint (pt) = 2 cups (c)
1 gallon (gal) = 4 qt = 8 pt = 16c = 128 fl oz
1 yard square (yd^2) = 9 ft^2 = 1,296 in^2
1 pound (lb) = 16 Ounces (oz)
1 long ton (T) = 2,240 lbs

1 yard (yd) = 3 feet = 36 inches (in)
1 acre = 43,560 square feet (ft^2)
1 cup (c) = 16 T = 8 fluid ounces (fl oz)
1 quart (qt) = 2 pt = 4c = 32 fl oz
1 foot square (ft^2) = 144 inch square (in^2)
1 acre = 4,840 yd
1 short ton = 2,000 pounds (lbs)

MARINER'S MEASURE:
1 fathom = 6 feet & 1 nautical mile = 1,013.4 fathoms

METRIC SYSTEM
1 meter (m) = 100 centimeters (cm)
1 Kilometer (Km) = 1,000 m
1 square meter (m^2) = 10,000 square centimeters (cm^2)
1 hectare (ha) = 10,000 square meters (m^2)
1 "manzana" = 0.698896 hectares = 7000 square meters
1 square Kilometer (Km^2) = 100 hectares (ha)
1 Kilogram (Kg) = 1,000 grams (g)
1 metric ton (t) = 1,000 Kg

CONVERSION OF UNITS between the two SYSTEMS
(Greatest unit mentioned first)
1 inch (in) = 2.54 centimeters (cm)
1 foot (ft) = 30 centimeters (cm) = 0.304 meters
1 meter (m) = 1.0936 yards (yd) = 3.2808 feet (ft)
1 mile = 1.609 Kilometers or 1 Kilometer = 0.6214 miles
1 square meter (m^2) = 10.76 square feet (ft^2) or 1 square foot = 0.093 sq. mts.
1 "manzana" = 1.73 acres
1 hectare = 2.47 acres = 107,600 sq. feet or 1 acre = 0.405 hectares
1 square mile (mi^2) = 2.59 square Kilometers (Km^2)
1 ounce (oz) = 28.35 grams (g)
1 Kilogram (Kg) = 2.2046 pounds (lb)
1 metric ton (t) = 1.1023 Tons (T)
1 fluid ounce (fl oz) = 29.575 milliliters (ml)
1 gallon (gal) = 3.785 liters (l)

TEMPERATURE CONVERSIONS
To convert FARENHEIT degrees into CELSIUS:
Subtract 32, multiply by 5 and divide by 9
To convert CELSIUS into FARENHEIT:
Multiply by 9, divide by 5 and add 32

THE CONVERSION OF UNITS MACHINE

The standard system that we use in the USA was created in England and the metric system that is used in many countries was created in France.

The conversion of units is used many times for solving math problems. See how "the conversion of units machine" works; it will help you do conversions within the standard system and also between the standard and metric systems.

Known unit → Question unit ?

96 inches → 1 foot = 12 inches   96 ÷ 12 = 8 → 8 Feet

Question: How many feet do we have in 96 inches?
The known unit INCH is SMALLER THAN the question unit FOOT
that is why we DIVIDE
Here is a rule to remember; if the known unit is smaller than the question unit you DIVIDE

Known Unit — FORMULA that relates the known unit with the question unit. See your chart. — Question unit ?

2 feet → 1 foot = 12 inches   2 (12) = 24 → 24 inches

Question: How many inches do we have in 2 feet?
The known unit FEET is GREATER THAN the question unit INCH
that is why we MULTIPLY
Here is a rule to remember; if the known unit is greater than the question unit you MULTIPLY

Create your own machine for solving many more problems!

Using the PROTRACTOR to draw REGULAR POLYGONS

Based on the circle, divide 360º by the number of sides of each regular polygon; you will find the angles for drawing each polygon

TRIANGLE 360º÷3=120º
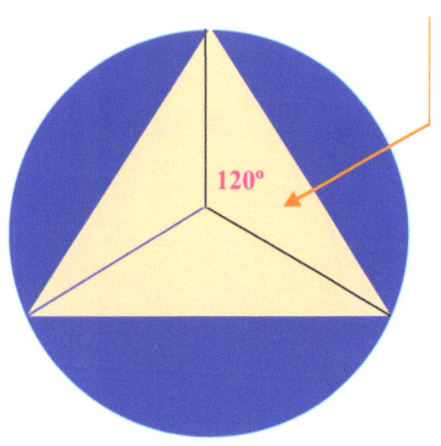

SQUARE 360º÷4=90º
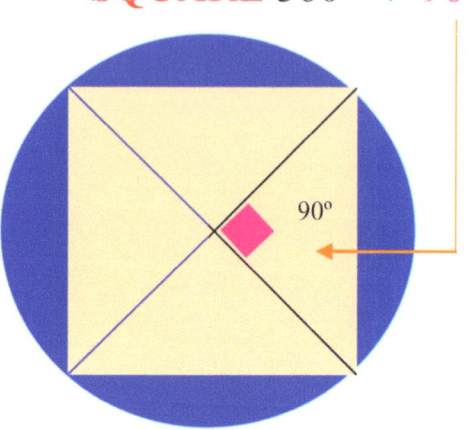

PENTAGON 360º÷5=72º
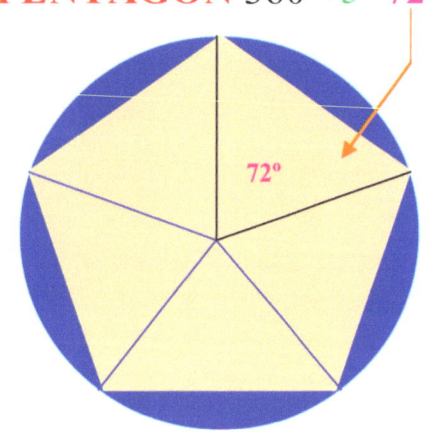

HEXAGON 360º÷6=60º
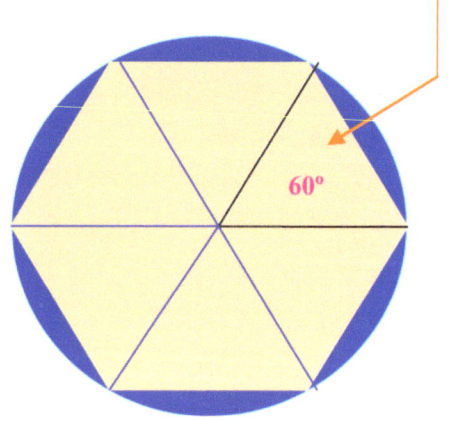

HEPTAGON 360º÷7≈51º
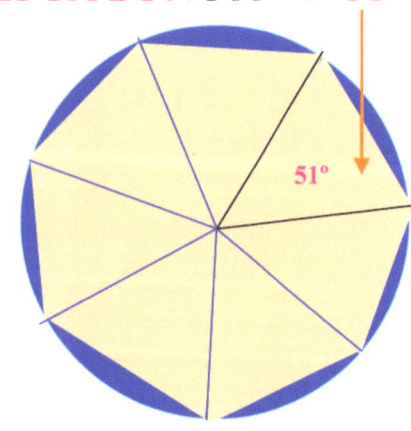

OCTAGON 360º÷8=45º
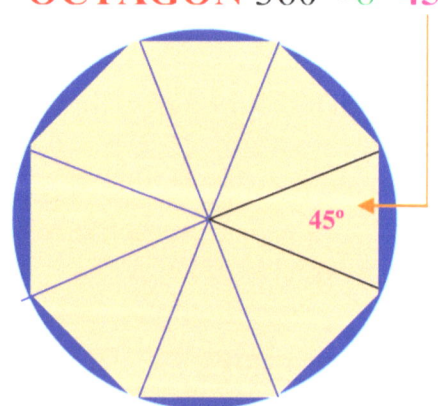

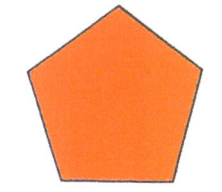

THE PENTAGON
and the PUMPKIN

Draw a REGULAR PENTAGON from a CIRCLE.
Fold the areas outside of the PENTAGON.

Use the PENTAGON as a template to draw 11 more PENTAGONS out of the same size CIRCLE...
Continue to fold the areas outside of the PENTAGONS.

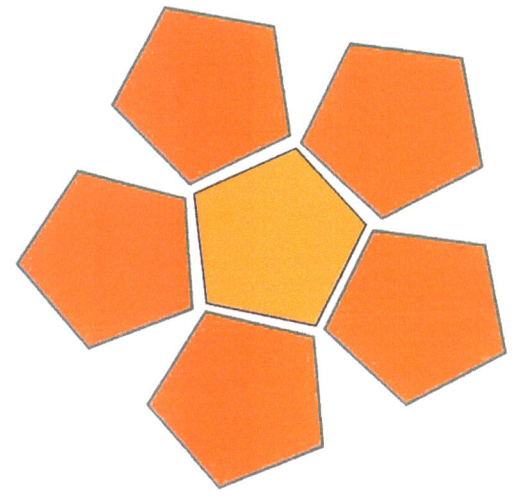

Put together 6 PENTAGONS by using the folded areas of the CIRCLE.

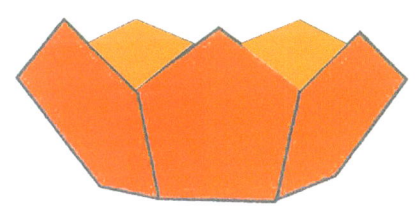

The 6 PENTAGONS together will take the shape of a basket.

This is half of the PUMPKIN.

We need the other half to complete the project.

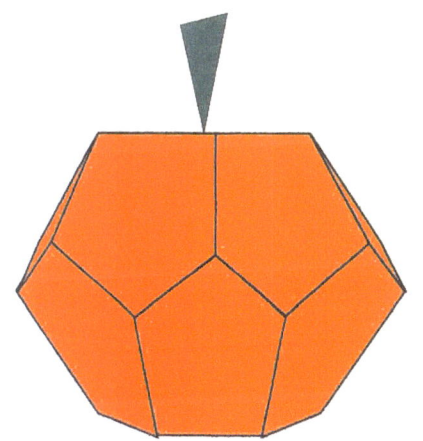

Put together the 2 halves and create a **DODECAHEDRON** and a **PUMPKIN.**

Discovering the FORMULA of the RECTANGLE

TOP
LENGTH

1	2	3	4	5	6	7	8
9	10	11	12	13	14	15	16
17	18	19	20	21	22	23	24
25	26	27	28	29	30	31	32

LEFT SIDE
WIDTH

RIGHT SIDE
HEIGHT

BASE

RECTANGLE

4 sides: 2 sides are called **BASE** or **LENGTH**
2 sides are called **HEIGHT** or **WIDTH**

4 right angles (90 degrees)

We can find the AREA by counting the squares or by multiplying one side by the other.

AREA = BASE (HEIGHT)   **A = bh**

Or   AREA = LENGTH (WIDTH)   **A = lw**

You already know the area of the rectangle and the FORMULA to find it…

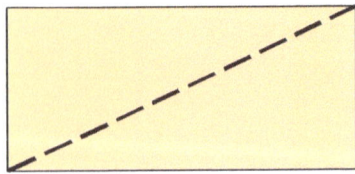

… What if we cut the rectangle in two, going from one VERTEX (corner) to the OPPOSITE VERTEX…?

Discovering the FORMULA of the TRIANGLE

We have a new POLYGON… THE TRIANGLE

We only have 3 sides.

We have 3 angles but only one of them is a right angle of 90 degrees

The AREA is going to be half of the area of the rectangle so we can write the formula of the TRIANGLE as BASE multiplied by HEIGHT but divided by 2, since it is the half of the rectangle

$$A = \frac{bh}{2}$$

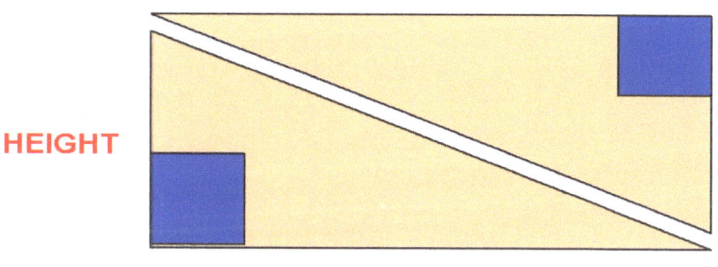

HEIGHT

BASE

All the TRIANGLES are HALF of a QUADRILATERAL…
We are going to discover the FORMULAS for all the different QUADRILATERALS.

Look for the PERPENDICULAR LINE. That is the HEIGHT…
It forms a right angle with the BASE.

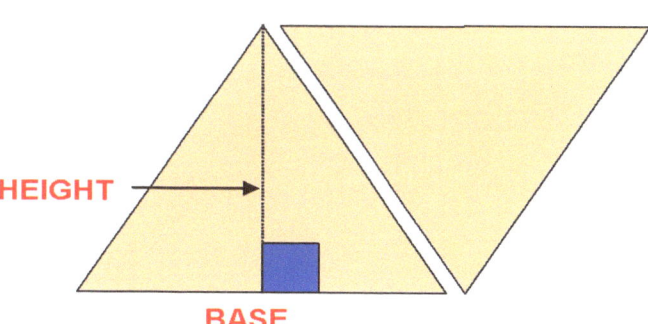

HEIGHT

BASE

Discovering the FORMULA of the PARALLELOGRAM

RECTANGLE		PARALLELOGRAM
	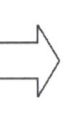	

If we cut one triangle on the right side of my rectangle and glue it to the left side, then I get another quadrilateral known as a …

PARALLELOGRAM

The base is parallel to the top and the right side is parallel to the left

The base is the same as the rectangle

The height is no longer the left or the right side

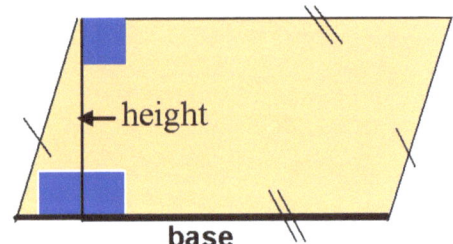

Remember…If I can follow the position of my right angles I can find my height!

My area is the same as the rectangle

And my formula for the area is the same

A = bh

The rectangle is also a parallelogram because it has two pair of sides that are parallel to each other. The difference between the rectangle and the parallelogram is that the rectangle has four right (90º) angles. This is why the parallelogram is not a rectangle

Discovering the FORMULA of the TRAPEZOID

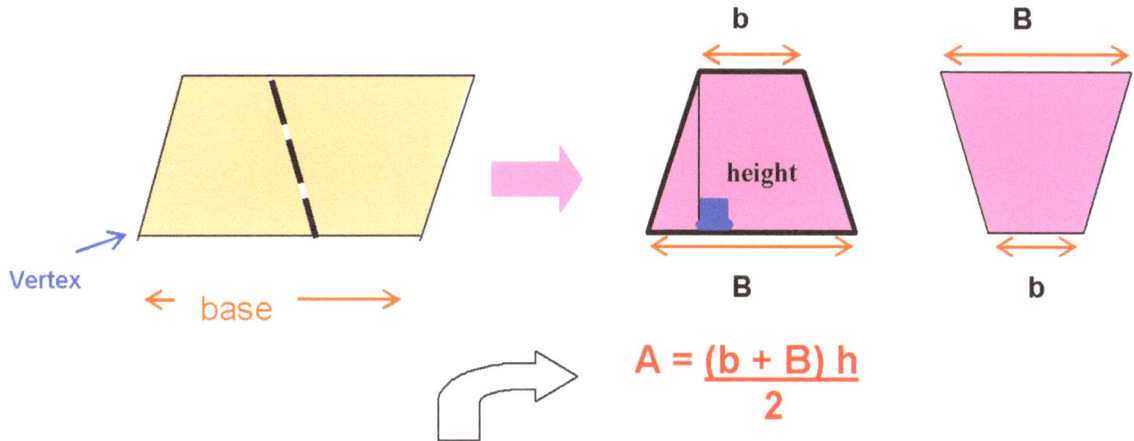

$$A = \frac{(b + B)h}{2}$$

Formula of the TRAPEZOID

If I have a parallelogram and I draw a diagonal line from top to bottom that goes through none of the **vertexes**, then I create a...

PARALLELOGRAM ➡ TRAPEZOID

If I put 2 of the same trapezoids together inverse to each other, then I have a:

PARALLELOGRAM

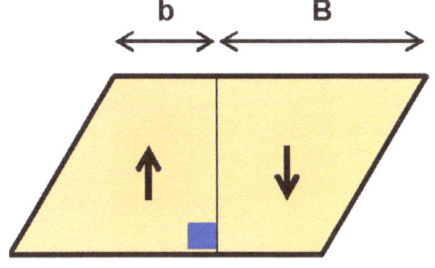

So if I can find the area of the parallelogram and divide it by 2, I can find the area of one trapezoid...

I just need to figure out my base and my height...

Area of TRAPEZOID = $\dfrac{\text{Area of PARALLELOGRAM}}{2}$

TRIANGLES
Classifying them by their sides & Classifying them by their angles:

The sum of the interior angles of any triangle is 180º

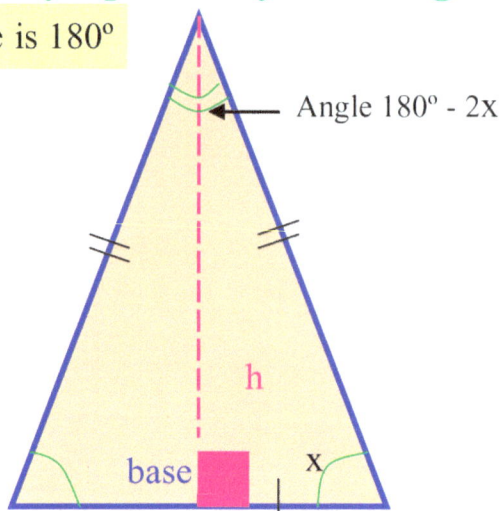

Angle 180º - 2x

EQUILATERAL
3 congruent sides (same size)
ACUTE
3 acute angles of 60º
3 congruent angles

ISOSCELES
2 congruent sides
ACUTE
3 acute angles (less than 90º)
2 congruent angles

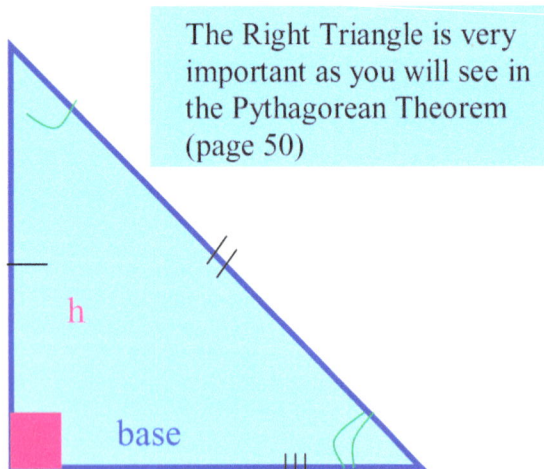

The Right Triangle is very important as you will see in the Pythagorean Theorem (page 50)

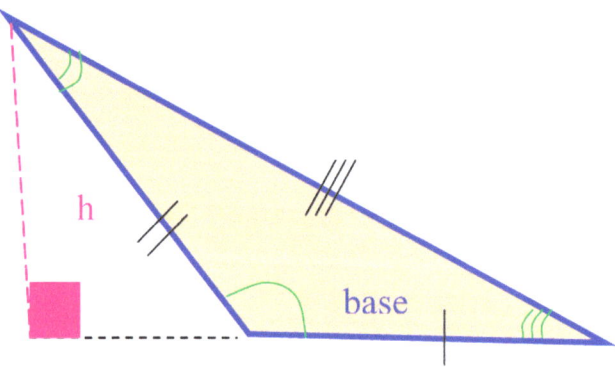

RIGHT TRIANGLE
1 Right angle (90º)
It can have 2 congruent sides
or the 3 sides can be different

SCALENE
It can have 2 congruent sides
or 3 different sides
OBTUSE one angle greater than 90º

Project: Using ruler, protractor and scissors cut and paste the different triangles on your note book.

It is very important to include the square that represents the 90º angle because it connects the base with the height "h"

PYTHAGOREAN THEOREM

$$a^2 + b^2 = c^2 \quad \text{and} \quad c = \sqrt{a^2 + b^2}$$

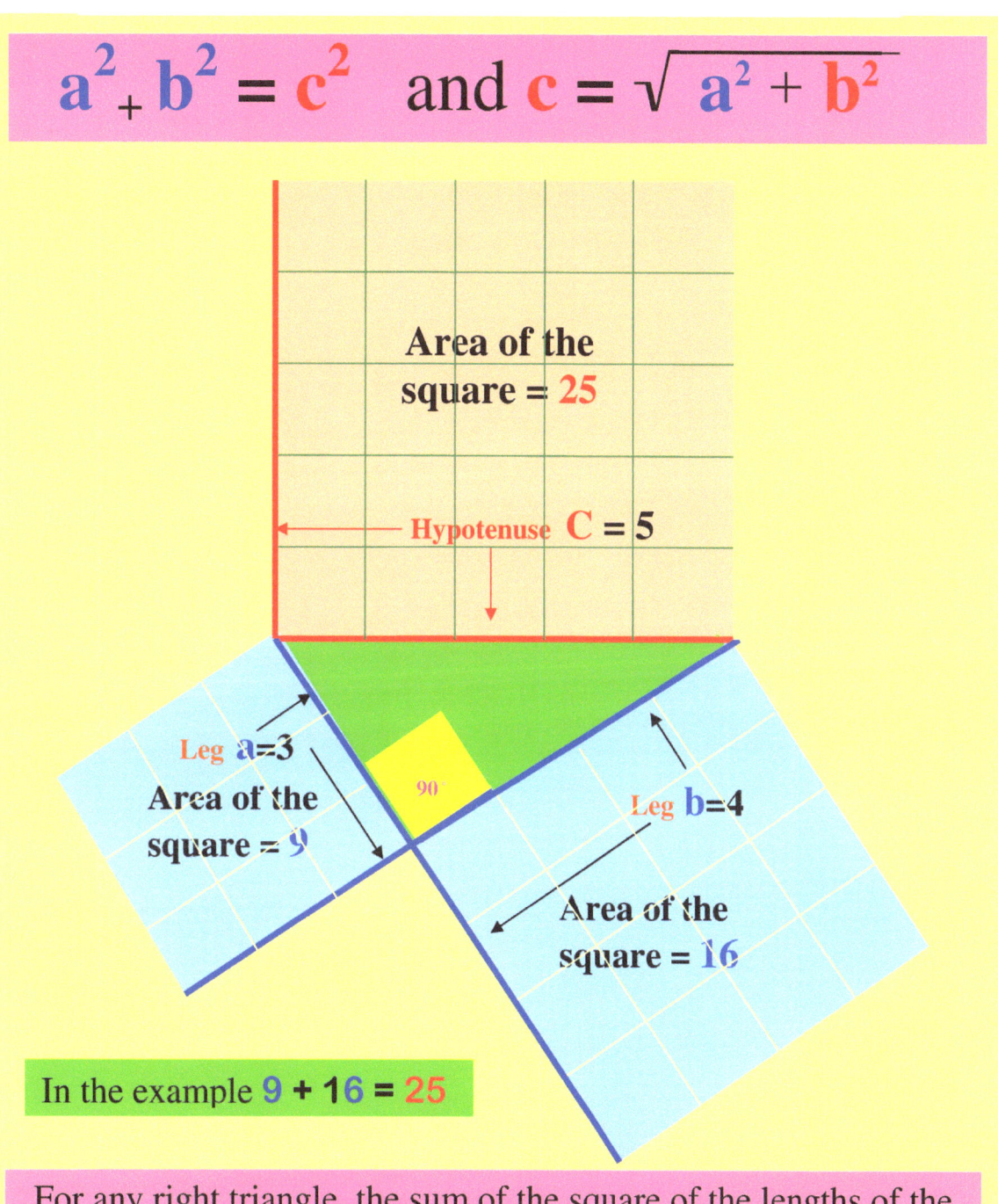

In the example 9 + 16 = 25

For any right triangle, the sum of the square of the lengths of the legs: **a** and **b**, equals the square of the length of the hypotenuse **c**

CIRCLE

Cut out the circumference of 4 circles in different colors with the following radiuses: 1 ¾", 1 ½", 3 centimeters and 1.9 centimeters. Fold them in half to find the diameter and if you fold them in half one more time you will find the radius.

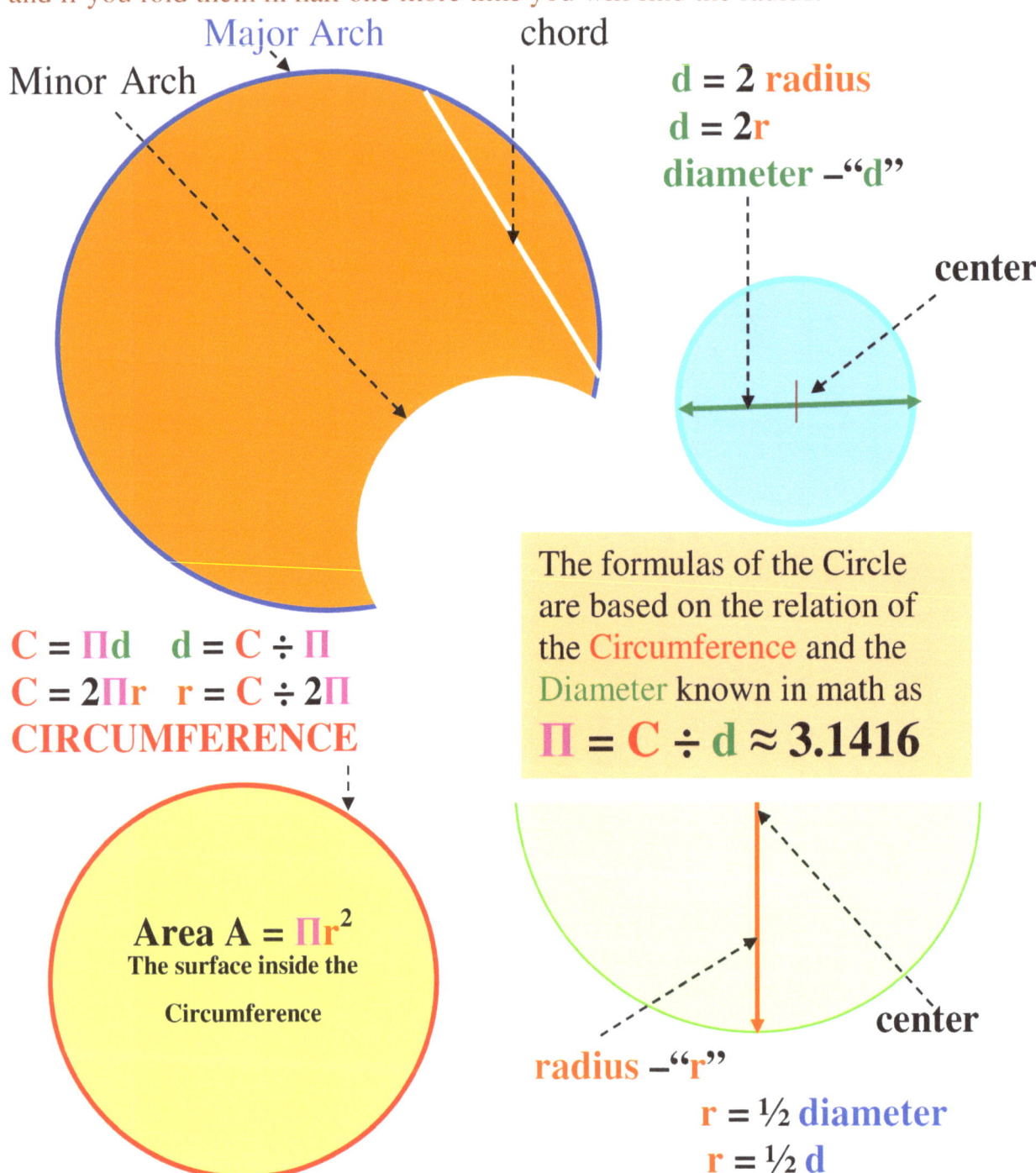

Major Arch
Minor Arch
chord

d = 2 radius
d = 2r
diameter – "d"

center

C = Πd   d = C ÷ Π
C = 2Πr   r = C ÷ 2Π
CIRCUMFERENCE

The formulas of the Circle are based on the relation of the Circumference and the Diameter known in math as
Π = C ÷ d ≈ 3.1416

Area A = Πr²
The surface inside the Circumference

radius – "r"
center
r = ½ diameter
r = ½ d

Cut your circles like the examples above and paste them on a paper for your notes…you already have the circle in your hands!

SURFACE AREA
Surface Area is the sum of all the faces of any volumetric figure

h=6 cm
w=2 cm
l=8 cm

Surface Area -"S. A." of a Rectangular Prism is the SUM of the areas of all the different faces:
- The bottom & the top
- The left & right faces
- The front & back faces

Remember your area formulas:
SQUARE is $A = s^2$
RECTANGLE A = lw or **bh**
TRIANGLE is $A = bh \div 2$
CIRCLE is $A = \pi r^2$

Length l=8

Bottom & Top   Width w=2

Area of Bottom = 2(8) =16 cm
Area of Top = 2(8) =16 cm

Left & Right face

Height h=6

Width w=2

Area of Left face = 2(6) =12 cm
Area of Right face = 2(6) =12 cm

Front & Back face

Height h=6

Area of Front face = 8(6) =48 cm
Area of Back face = 8(6) =48 cm

Length l=8

SURFACE AREA= 16+16+12+12+48+48=152 cm

Can you find the surface area of a cereal box?

VOLUME
OF PRISMS & CYLINDER
The Area of the Base multiplied by the height "H"

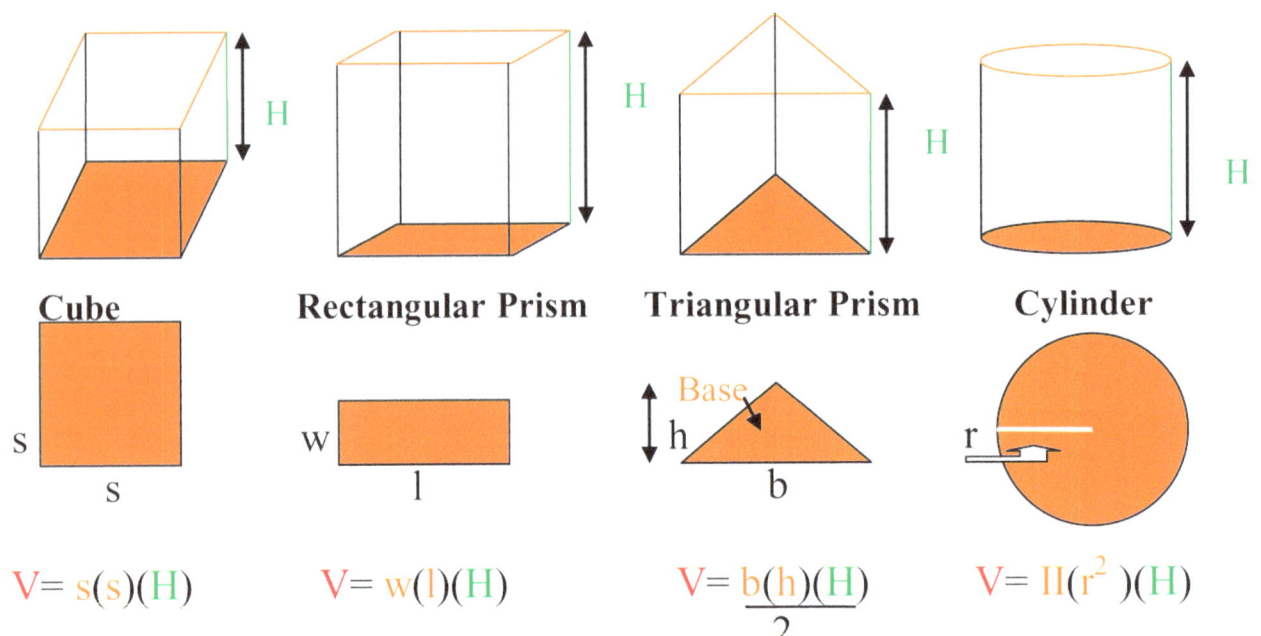

Cube	Rectangular Prism	Triangular Prism	Cylinder
$V = s(s)(H)$	$V = w(l)(H)$	$V = \dfrac{b(h)(H)}{2}$	$V = \Pi(r^2)(H)$

VOLUME
OF PYRAMIDS & CONE

⅓ of the Area of the Base multiplied by the height "H"
See the page of Cylinder VS Cone to understand why 1/3

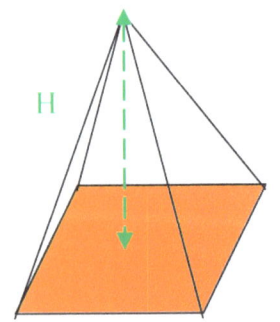
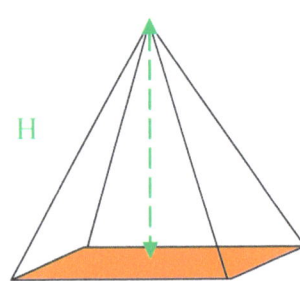
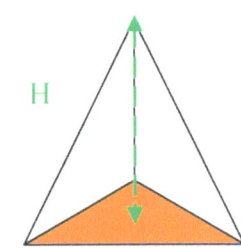
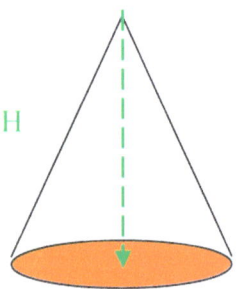

Square Pyramid (Square Base)
$V = ⅓ (s)(s)(H)$

Rectangular Pyramid (Rectangular Base)
$V = ⅓ (w)(l)(H)$

Triangular Pyramid (Triangular Base)
$V = ⅓ \dfrac{(b)(h)(H)}{2}$

Cone (Circular Base)
$V = ⅓ \Pi(r^2)(H)$

CYLINDER

Template to build one CYLINDER with the Base and Height of a commercial CONE size.
Dimensions:
Radius = 4.1 cm.
Face height = 10.3 cm.
Face length = 25.7 cm.
SURFACE AREA
Add the areas:

$$\begin{aligned}\text{Top} &= 52.78 \text{ cm}^2 \\ \text{Base} &= 52.78 \text{ cm}^2 \\ \text{Face} &= 264.71 \text{ cm}^2\end{aligned}$$

Surface area = 370.27 cm²

AREA $A = \Pi r^2$
A=3.14 (4.1) (4.1) = 52.78cm²

Radius = 4.1 cm

Diameter d = 8.2 cm

Circumference $C = \Pi d$
C= 3.14(8.2) = 25.7 cm

Face of the Cylinder:
Length = Circumference of the Base = 25.7 cm
Height = 10.3 cm
(The height needs to be the same height of a commercial cone that will be used for the Project "Cone VS Cylinder")

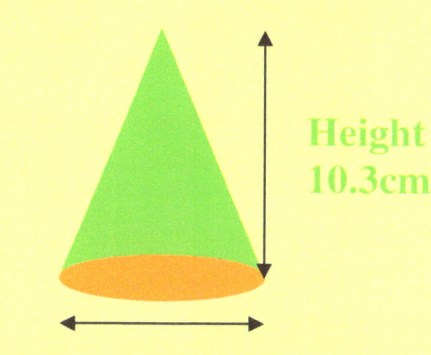

Height 10.3cm

Diameter 8.2cm
Circumference 25.7cm

VOLUME OF THE CYLINDER

V= AREA of the base (height)
V = 52.78 cm² (10.3 cm)
V = 543.63 cm³

CYLINDER VS CONE

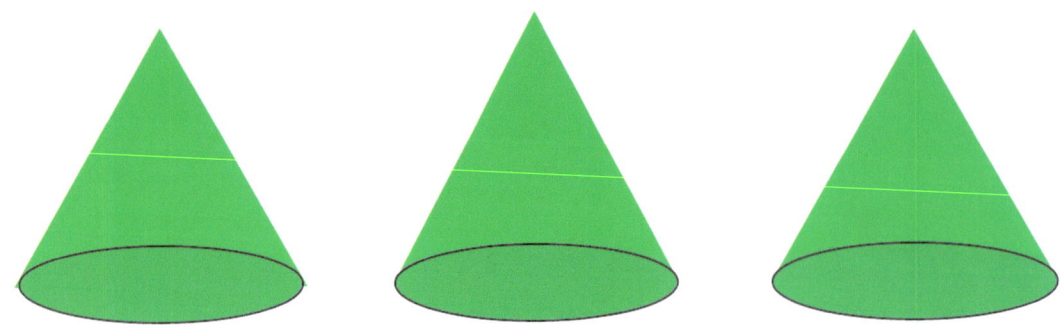

The VOLUME of 3 CONES = The VOLUME of 1 CYLINDER
(When their base and height is the same)

VOLUME of the CONE = $\dfrac{1}{3}$ VOLUME of the CYLINDER.

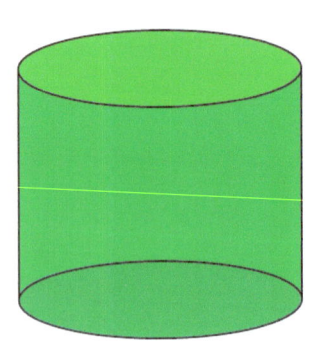

EXPERIMENT: We are going to build a CYLINDER with the same BASE and the same HEIGHT of a commercial CONE. We are going to fill the 3 CONES with rice and empty them into the CYLINDER.

Experiment: Cut the CONE as shown in the drawing and you will discover that the SURFACE AREA of the face is ONE THIRD of the area of a CIRCLE.

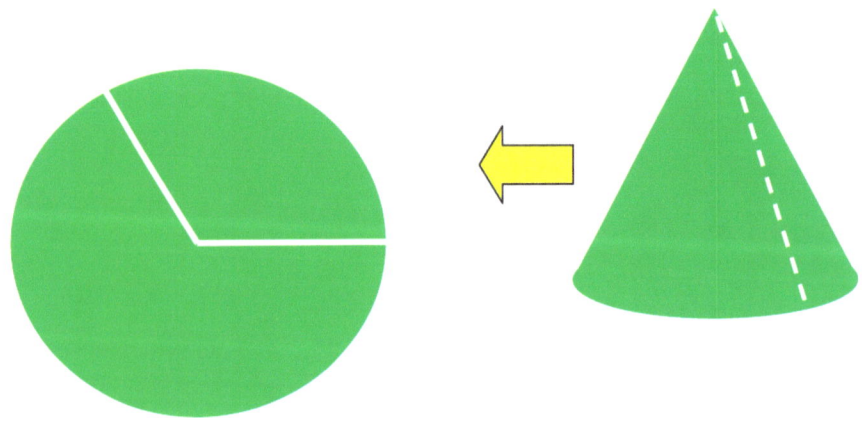

THE MOBILE

Cut the following geometric figures out of construction paper and build a MOBILE in BALANCE. Start with a rectangle 4" by 8" that has an area of 32 square inches:

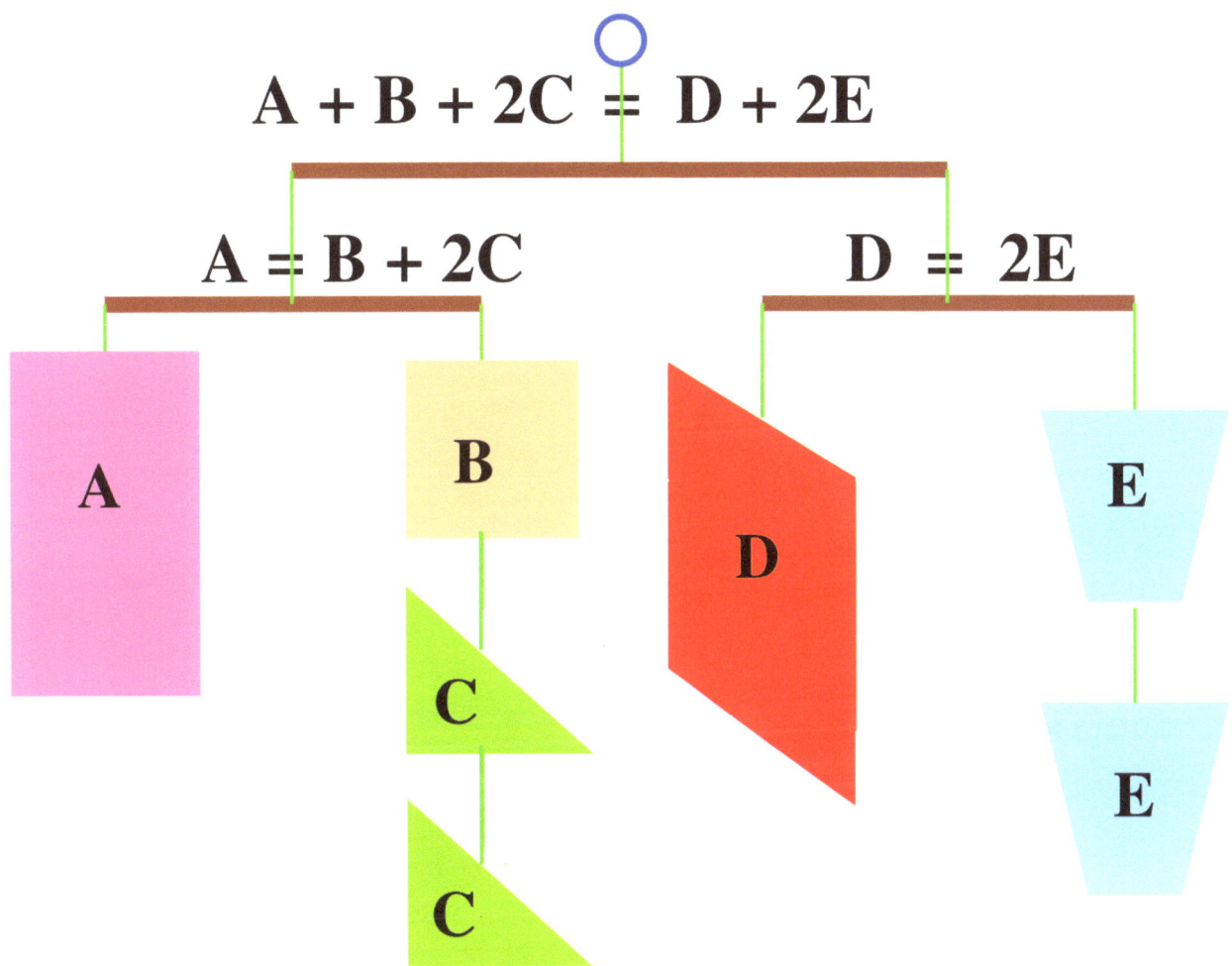

THE AREAS OF THE DIFFERENT POLYGONS ARE IN BALANCE.

A = B + 2C

The area of the RECTANGLE = the area of the SQUARE
+ the area of the 2 TRIANGLES.

D = 2E

The area of the PARALLELOGRAM = the area of 2 TRAPEZOIDS.

A + B + 2C = D + 2E

The areas of the RECTANGLE + SQUARE + 2 TRIANGLES is equal to the areas of the PARALLELOGRAM + 2 TRAPEZOIDS.

THE LANGUAGE OF ALGEBRA and the ghost

Algebra is a branch of mathematics that includes the use of variables to express rules about numbers, number relationships and operations.

In algebra we write numbers and letters
GHOST…some times we do not write numbers or signs

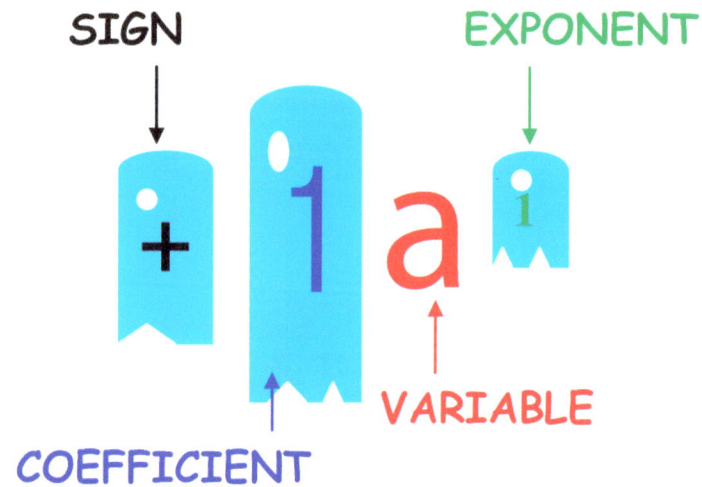

Coefficient is the numerical part of a variable term
Variable is a letter that represents one or more numbers

Other GHOSTS... $\underline{6}$, - 5(a) , x(y) , a(b)(c)(d)

The denominator 1   The multiplication sign between:

- A coefficient and a variable
- Between variables

Algebraic expression: An expression that consists of numbers and variables and operations to be performed.

Example: $3x + 2y^3 - 5$ ← **Constant** is a number with no variable

The Mobile and ADDITION EQUATIONS

Before you start solving an equation, you can rotate it so that the variable is in the left side. In the mobile you can see that the balance is the same if you have:

$5 = a + 2$ or $a + 2 = 5$

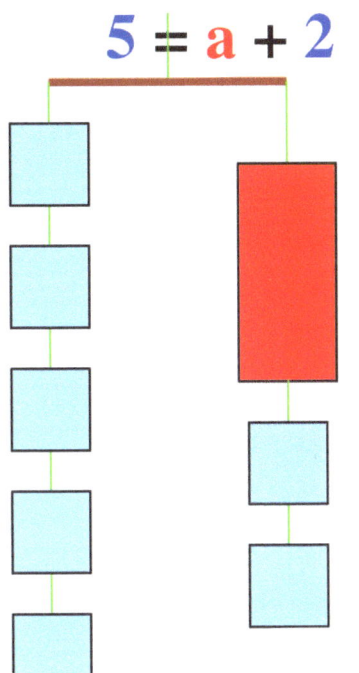
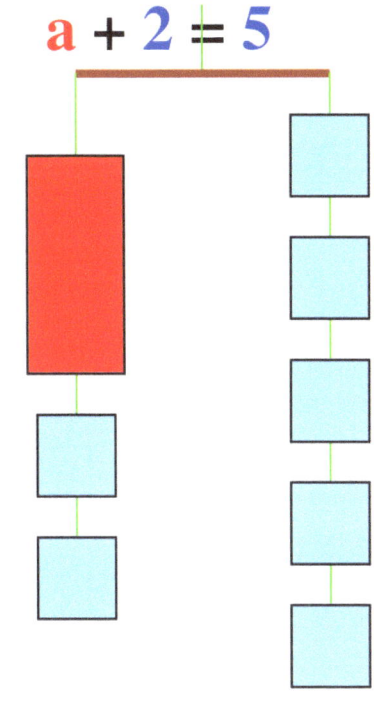

Solving the Equation means finding the value of the variable.
If you take away two units from each side, you will discover that the mobile is in balance! And the value of the variable is equal to three units

$a + 2 = 5$
$-2 \quad -2$
$a = 3$

Here is a "tip" for solving the equation that is going to be great for solving very complex equations in the future:
 Think of the equal sign as the line of balance.
 When solving the equation every time you cross the line of balance do the opposite operation.

Look: $a + 2 = 5$
$a = 5 - 2$
$a = 3$

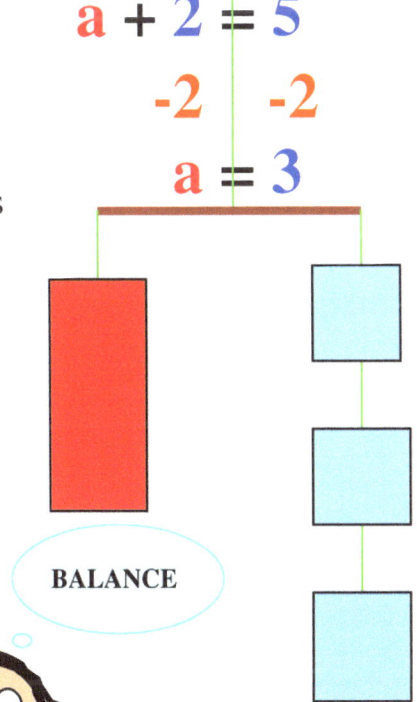

BALANCE

58

The Mobile and MULTIPLICATION EQUATIONS

The beauty of the mobile is that it helps us understand why is that equations are mathematical expressions in balance. See how it works with multiplication:

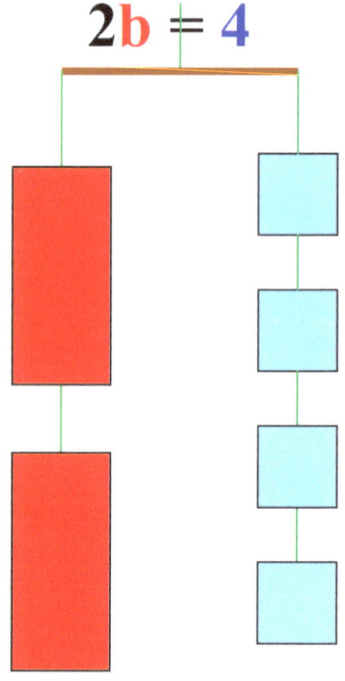

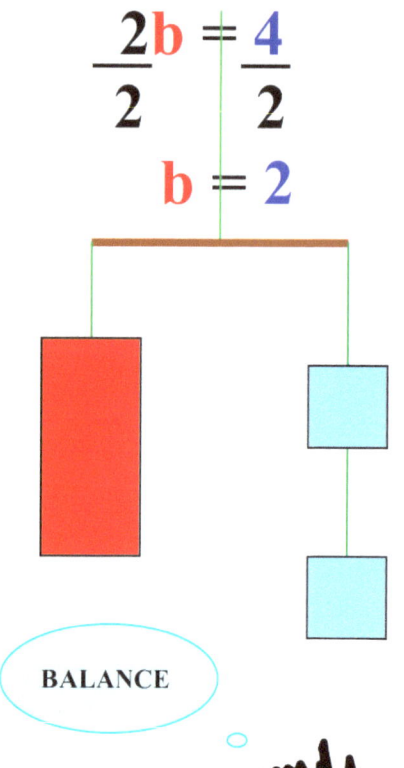

BALANCE

Solving the Equation means finding the value of one variable.
If you take away half of the units from each side, you will discover that the mobile is in balance! And the value of the variable is equal to two units

Use the same system for solving the multiplication equation:
  Think of the equal sign as the line of balance.
  When solving the equation every time you cross the line of balance do the opposite operation.
  The opposite operation of multiplication is division

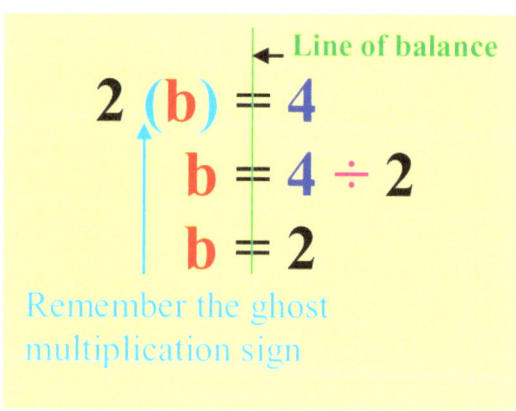

Remember the ghost multiplication sign

We are going to continue solving equations by doing opposite operations. Look at the examples on next page to see how we solve simple equations of addition, subtraction multiplication and division. Later on look at the colorful two step equations using the same system. It works great!

59

Solving simple EQUATIONS

An Equation is a statement of equality between two quantities

After looking at the "mobile" equation; we are going to follow a system to solve equations by just remembering to do the opposite operation every time you cross the line of balance of the equation.

addition

$$a + 2 = 4$$
$$a = 4 - 2$$
$$a = 2$$

The opposite operation of addition is subtraction.

subtraction

$$b - 3 = 5$$
$$b = 5 + 3$$
$$b = 8$$

The opposite operation of subtraction is addition.

multiplication

$$2c = 4$$
$$2(c) = 4$$
$$c = \frac{4}{2}$$
$$c = 2$$

ghost ↗

The opposite operation of multiplication is division.

division

$$\frac{d}{3} = 6$$
$$d = 6\,(3)$$
$$d = 18$$

The opposite operation of division is multiplication.

Remember to rotate the equation when needed so that the variable is on the left side of the equation. Look at your identical equations "mobile".

60

SOLVING INEQUALITIES

If the sign beside the coefficient is positive, solve the inequality like any simple equation.
 (Remember the ghost positive sign and the ghost coefficient 1)
Examples:

$+1a + 6 > 12$
$a > 12 - 6$
$a > 6$

a is greater than 6

$+1b - 4 < 9$
$b < 9 + 4$
$b < 13$

b is smaller than 13

$+2c \leq 8$
$c \leq 8 \div 2$
$c \leq 4$

c is equal or smaller than 4

$x \div 3 \geq 7$
$x \geq 7\,(3)$
$x \geq 21$

x is equal or greater than 21

If the sign beside the coefficient is negative, when you are solving the equation you multiply or divide by a negative number, and you must change the direction of the INEQUALITY. See the examples:

$-2m \geq 8$
$m \geq 8 \div (-2)$
$m \geq -4$

m is equal or greater than −4

$y \div -3 > 12$
$y < 12\,(-3)$
$y < -36$

y is smaller than -36

WRITING ALGEBRAIC EXPRESSIONS

Algebraic Expressions are "Variables" (letters) and "Constants" (numbers) doing operations...remember that the numerical Factor of a variable term is the "Coefficient"
Real world problems can be written as algebraic expressions
Examples:
AT&T has a plan for the cost of a cellular phone that offers a flat fee of $20 plus 4 cents per minute. You write: $0.04x + 20$
The formula for the volume of a rectangular pyramid is one third of the area of the base multiplied by the height. You write: $V = \frac{1}{3} lwh$

We follow some rules when we write Algebraic Expressions:
- We write in alphabetical order
$$2a + 5b - 3c$$
- We write from largest to smallest exponent of like variables
$$5a^3 + 2a^2 + 3a$$
- The "Constants" (numbers without variables) goes last
$$7a^8 + 5a^4 + 2b^{14} - 8$$

Algebraic Expressions have names depending on the number of **"Terms"**.

Terms are separated by $+$ or $-$ signs
MONOMIAL = one term
5 or xy or $7ab^3$
BINOMIAL = two terms
$a + 5$ or $x - y$ or $7ab^3c^2 - 3fg$ or $a^9 - a^2$
TRINOMIAL = three terms
$a^2 + 2ab + b^2$
POLYNOMIAL = A sum of a finite number of monomials
$x^3 - 5xy + y$ or $a^4 + 3a^2b - 3ab^2 - b^7$

EXPRESSIONS AND EQUATIONS
The difference between an expression and an equation is that the Equation has an answer
$5x + 7$ is an Algebraic Expression and $5x + 7 = 12$ is an Equation
Note: you are going to learn how to solve Algebraic Expressions and Equations

ADDITION IN ALGEBRA - Rules & Examples

You can only add like terms...
- Same alphabetical variables
- with the same exponent

$1a^3 + 1a^3 = 2a^3$

You never add different variables

$1x + 1y = x + y$

You never add variables with numbers

$1b + 5 = b + 5$

The color code is going to help you do your addition problems: Black for the signs, Blue for the Coefficients, Red for the variables and green for the exponents

How to do it...
SIGNS:
- same rules as addition of integers
- remember the + ghost sign when needed

COEFFICIENTS:
- add only the like terms
- remember the 1 ghost coefficient when needed

VARIABLES:
- Stay the same. DO NOT CHANGE THE EXPONENTS!
- Remember to order your answer in alphabetical order starting with the largest exponent

$$+2a + 3a = +5a$$

$-4xy^5 - 6xy^5 = -10xy^5$

$+5y^2 + 3x^4 - 2y^2 - 7x^4 = -4x^4 + 3y^2$

$-8ab^7 + 6ab^7 = -2ab^7$

$-3a^6 + 8b^9 - 5b^9 - 9 + 1a^6 = -2a^6 + 3b^9 - 9$

MULTIPLICATION IN ALGEBRA-Rules & Examples

You can multiply anything in algebra …
- Same or different alphabetical variables
- with the same or different exponent

Remember:
Black for the signs
Blue for Coefficients
Red for Variables
Green for the exponents

How to do it…
SIGNS:
- same rules as multiplication of integers
- remember the + ghost sign when needed

COEFFICIENTS:
- multiply
- remember the 1 ghost coefficient when needed

VARIABLES:
- ADD THE EXPONENTS of the same alphabetical variables
- remember to order your answer in alphabetical order
- starting with the largest exponent

$a \cdot a \cdot a = +1a^1 \cdot +1a^1 \cdot +1a^1 = +1a^3 = a^3$

Add the exponents of the like variables
Multiply the Coefficients

$-4x^7y^2(-6x^2y^4) = +24x^9y^6$

$-8a^1b^6(+6a^2b^4) = -48a^3b^{10}$

$-3a^5(-5b^4-9)$ means $-3a^5(-5b^4)$ and $-3a^5(-9) = +15a^5b^4 + 27a^5$

Strategy for multiplying a binomial by a binomial: multiply the 1st term of the 1st binomial by the two terms of the 2nd binomial, than multiply the 2nd term of the 1st binomial by the two terms of the 2nd binomial

$(+2y^2+3)(+4x^4-3y^1) = +8x^4y^2 - 6y^3 + 12x^4 - 9y^1$

DIVISION IN ALGEBRA - Rules & Examples

You can divide anything in algebra ...
- Same or different alphabetical **variables**
- with the same or different **exponent**

Follow the color:
Black for signs
Blue for Coefficients
Red for variables
Green for exponents

How to do it...
SIGNS:
- same rules as division of integers
- remember the + ghost sign when needed

COEFFICIENTS:
- divide
- remember the 1 ghost coefficient when needed

VARIABLES:
- **SUBTRACT THE EXPONENTS** of the same alphabetical variables. Remember the ghost exponent 1
- remember to order your answer in alphabetical order
- starting with the largest exponent

$$+12a^8 \div +3a^6 = +4a^2$$

$-24x^7y^5 \div -8x^4y^2 = +3x^3y^3$

$+75y^6 \div -25y^{12} = -3y^{-6}$ or $\dfrac{-3}{y^6}$

$-8a^3b^8 \div +2a^3b^5 = -4b^3$

$-30a^9b^{12}c^6 \div +15a^4b^{12}c^2 = -2a^5c^4$

$+6x \div -3 = -2x$

$\dfrac{+21x^{10}y^{15} - 18x^9y}{3x^{10}y^{10}} = 7y^5 - 6x^{-1}y^{-9}$ or $7y - \dfrac{6}{xy^9}$

POWERS IN ALGEBRA-Rules & Examples

They are repeated multiplication written in exponential form...
$2·2·2·2 = 2^4$ $3a·3a·3a = (3a)^3$ $-5x^3(-5x^3) = (-5x^3)^2$

The ( ) indicate that the exponent applies to the complete expression

Black for signs, Blue for Coefficients, Red for Variables and Green for Exponents

How to do it...

SIGNS:
- If the sign of the algebraic term is + the answer is always +
- If the sign of the algebraic term is NEGATIVE there are two rules:

    EVEN exponent of the algebraic expression, the answer is +

    ODD exponent of the algebraic expression, the answer is -
- remember the + ghost sign when needed

COEFFICIENTS:
- Multiply the base coefficient as many times as described by the exponent
- remember the 1 ghost coefficient when needed

VARIABLES:
- Multiply the EXPONENTS of each variables by the EXPONENT of the algebraic term
- remember to order your answer in alphabetical order, starting with the largest exponent

$$(+12a^1)^2 = +144a^2$$

$(-2x^3y^4)^2 = +4x^6y^8$
$(-5a^6b^5)^3 = -125a^{18}b^{15}$

EXPONENTS – Rules and Examples:

$10^0 = 1$ $0.5^2 = 0.5 \cdot 0.5 = 0.25$

$10^1 = 10$ $(2/3)^2 = (2/3)(2/3) = 4/9$

$-10^2 = -(10 \cdot 10) = -100$

Negative base and no parentheses answer = – (negative)

Negative base inside of parentheses: Look at the exponent…

$(-10)^2 = (-10)(-10) = +100$ even exponent = +

$(-10)^3 = -10(-10)(-10) = -1000$ odd exp. = –

Negative exponents:

$10^{-1} = 1/10^1$ negative exponent $a^{-n} = 1/a^n$

$1/10^{-1} = 10^1$ denominator negative exponent $1/a^{-n} = a^n$

DOING OPERATIONS WITH EXPONENTS:

For multiplication add the exponents $a^m \cdot a^n = a^{m+n}$

$10^2 \cdot 10^3 = 10^{2+3=5} = 10^5$

$10^{-3} \cdot 10^5 = 10^{-3+5=2} = 10^2$

$10^{-4} \cdot 10^{-3} = 10^{-4+-3=-7} = 10^{-7} = 1/10^7$

For division subtract the exponents $a^m/a^n = a^{m-n}$

$10^4/10^2 = 10^{4-2=2} = 10^2$

$10^5/10^{-3} = 10^{5-(-3)=5+3=8} = 10^8$

$10^{-4}/10^{-2} = 10^{-4-(-2)=-4+2=-2} = 10^{-2} = 1/10^2$

For powers of powers multiply the exponents $(a^m)^n = a^{mn}$

$(10^3)^4 = 10^{3 \cdot 4=12} = 10^{12}$

TWO STEPS EQUATIONS with multiplication

VARIABLES	=	NUMBERS

← Line of balance

	=	
1 8	=	3 a + 6

Rotate the equation so that the variables are on the left side.

3 a + 6	=	1 8

Transfer the number to the side of the numbers and remember to do the opposite operation.

3 a	=	1 8 - 6

Do your operations of addition or subtraction.

3 (a)	=	1 2

Change the coefficient of the variable to the side of the numbers doing the opposite operation.

a	=	1 2 / 3

Remember there is a GHOST multiplication sign between a number and a variable.

Do your operations of division.

a	=	4

TWO STEPS EQUATIONS with division

| VARIABLES | = | NUMBERS |

← Line of balance

| 28 | = | X/-3 - 5 |

Rotate the equation so that the variables are on the left side.

| X/-3 - 5 | = | 28 |

Transfer the number to the side of the numbers and remember to do the opposite operation.

| X/-3 | = | 28 + 5 |

Do your operations of addition or subtraction.

| X/-3 | = | 33 |

Change the coefficient under the variable to the side of the numbers doing the opposite operation.

| X | = | 33 (-3) |

Remember to carry the sign of the denominator when you transfer it to the opposite side of the equation. Also follow the rules of the SIGNS for integers when you do your operations.

Do your operation of multiplication

| X | = | -99 |

TWO STEPS EQUATIONS with LIKE TERMS

| VARIABLES | = | NUMBERS |

← Line of balance

| $-9m + 11 + 2m + 6$ | = | $-3 - 8 + 7m$ |

Do all operations that you can on each side of the equation.

| $-7m + 17$ | = | $-11 + 7m$ |

Transfer the numbers to the side of the numbers and the variables to the side of the variables. Remember to do the opposite operation.

| $-7m - 7m$ | = | $-11 - 17$ |

Do your operations of addition or subtraction.

| $-14(m)$ | = | -28 |

Change the number besides the variable to the side of the numbers doing the opposite operation.

| m | = | $-28 / -14$ |

Remember there is a GHOST multiplication sign between a number and a variable, also follow the rules of the SIGNS for integers.

Do your operation of division.

| m | = | 2 |

TWO STEPS EQUATIONS with LIKE TERMS and FRACTIONS

VARIABLES	=	NUMBERS

← Line of balance

$b/4 - 3 - 9$	=	$-b/2 + 5 + 1$

Do all operations that you can on each side of the equation.

$b/4 - 12$	=	$-b/2 + 6$

Transfer the numbers to the side of the numbers and the variables to the side of the variables.
…Remember to do the opposite operation.

$b/4 + b/2$	=	$6 + 12$

Do your operations of addition or subtraction.

★ $3/4\ b$	=	18

Change the coefficient of the variable to the side of the numbers doing the opposite operation.

b	=	$18 \div 3/4$

→ ¾ (b) is the same as $3b/4$

REMEMBER there is a GHOST multiplication sign between the coefficient $3/4$ and the variable b. Also follow the rules of the signs for integers.

Do your operation of division $18 \div 3/4 = 18(4/3) = 72/3 = 24$

b	=	24

COORDINATE PLANE - Transformations

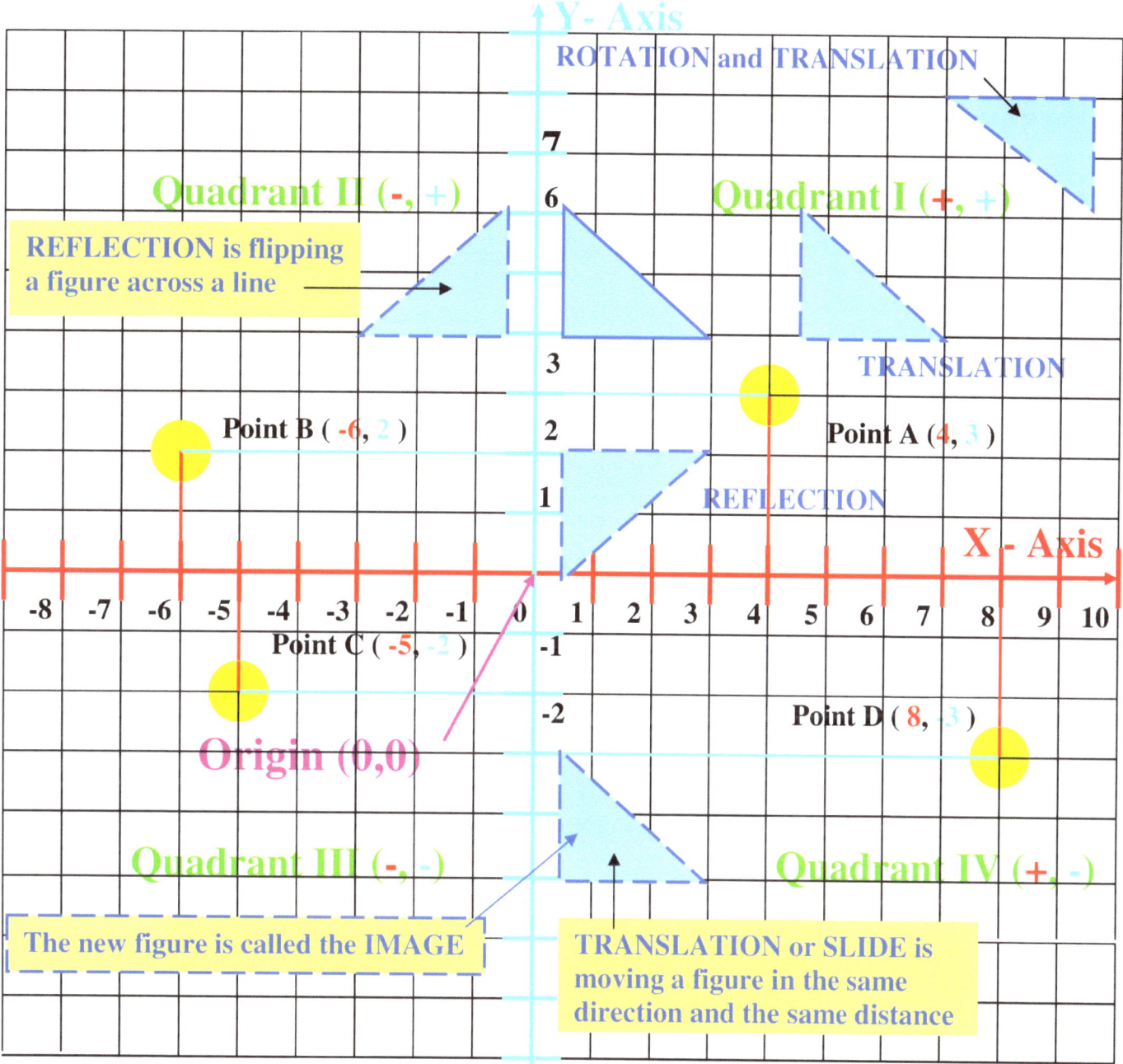

The Coordinate Plane help us to graph points. Each point has 2 coordinates (an ordered pair)
   Examples: Point **A** (4, 3)   Point **B** (-6, 2)
              Point **C** (-5, -2)   Point **D** (8, -3)
   The first coordinate is X, the second coordinate is Y
Understanding the coordinate plane is going to help us solve TRANSFORMATIONS problems: TRANSLATION, REFLECTION and ROTATION
Find the coordinates of the vertices of the different figures.

72

Linear Equations

Linear Equations are:
Equation with two variables to the first power that appear in separate terms.
They are Linear Functions because you can find their solution by choosing the value for one of the variables the "domain" to find the value of the other one, the "range".
In general they have more than one solution.
We can write them in three different forms:

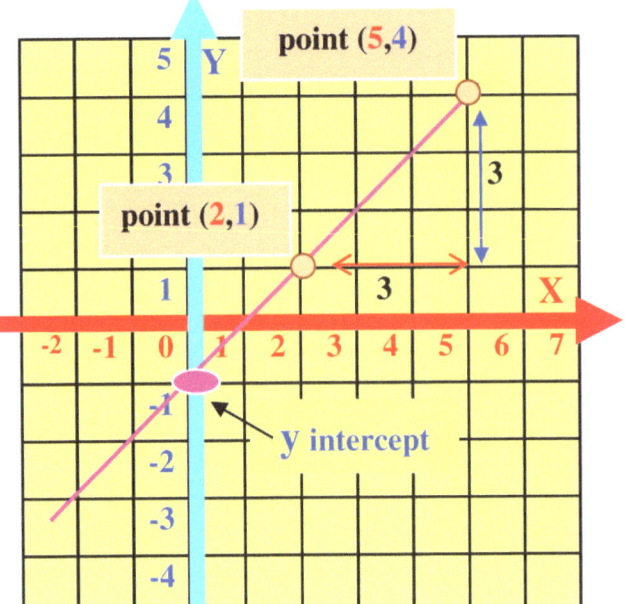

POINT-SLOPE FORM

$$y - y_1 = m(x - x_1)$$

(x_1, y_1) represent one point of the line … Example point (5,4)

m represents the SLOPE = $\dfrac{\text{Vertical change – RISE}}{\text{Horizontal change – RUN}} = \dfrac{3}{3}$

$m = \dfrac{y_2 - y_1}{x_2 - x_1}$ (x_1, y_1) and (x_2, y_2) are two points in the line

SLOPE – INTERCEPT FORM

$y = mx + b$ ← b represents the y-intercept

STANDARD FORM

$Ax + By = C$ $A > 0$ (is a positive number)

B and C are any integer

Notes: A line with $m = 0$ is a HORIZONTAL LINE

A line with m = undefined is a VERTICAL LINE

FUNCTIONS

A function is a relation (x,y) in which each element in the domain (x) is paired with exactly one element in the range (y)

The vertical line test can help you determine when a relation is a function; It should cross the graph only in one point (x,y)

Examples of some Linear and Power Functions and their graphs:

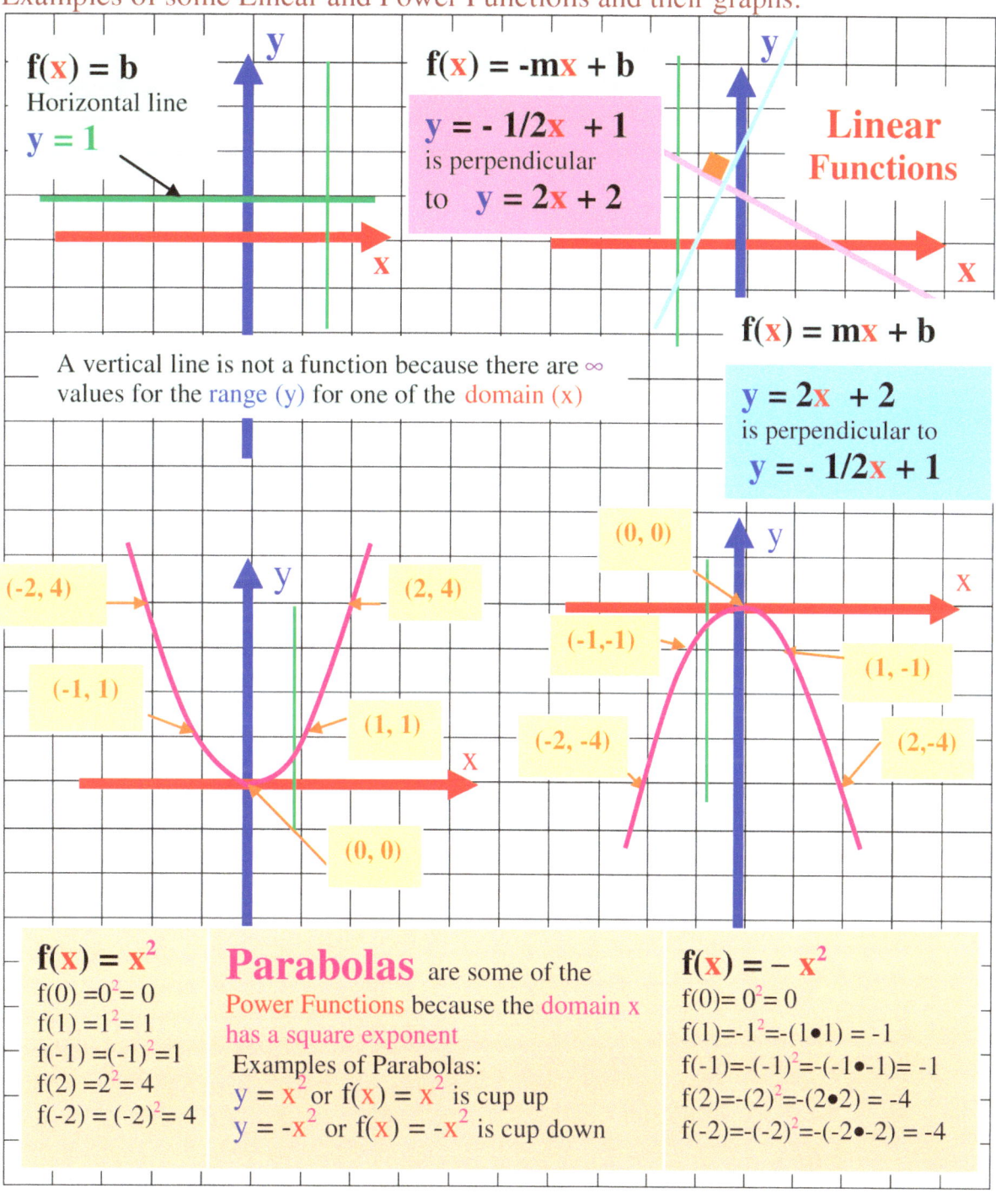

f(x) = b
Horizontal line
y = 1

f(x) = -mx + b
y = - 1/2x + 1 is perpendicular to y = 2x + 2

Linear Functions

A vertical line is not a function because there are ∞ values for the range (y) for one of the domain (x)

f(x) = mx + b
y = 2x + 2 is perpendicular to y = - 1/2x + 1

$f(x) = x^2$
$f(0) = 0^2 = 0$
$f(1) = 1^2 = 1$
$f(-1) = (-1)^2 = 1$
$f(2) = 2^2 = 4$
$f(-2) = (-2)^2 = 4$

Parabolas are some of the Power Functions because the domain x has a square exponent
Examples of Parabolas:
$y = x^2$ or $f(x) = x^2$ is cup up
$y = -x^2$ or $f(x) = -x^2$ is cup down

$f(x) = -x^2$
$f(0) = 0^2 = 0$
$f(1) = -1^2 = -(1 \cdot 1) = -1$
$f(-1) = -(-1)^2 = -(-1 \cdot -1) = -1$
$f(2) = -(2)^2 = -(2 \cdot 2) = -4$
$f(-2) = -(-2)^2 = -(-2 \cdot -2) = -4$

Discovering PATTERNS.

Patterns are the way numbers, things or objects change in sequence INCREASING or DECREASING…

FORMULA that describes the PATTERN

S E Q U E N C E

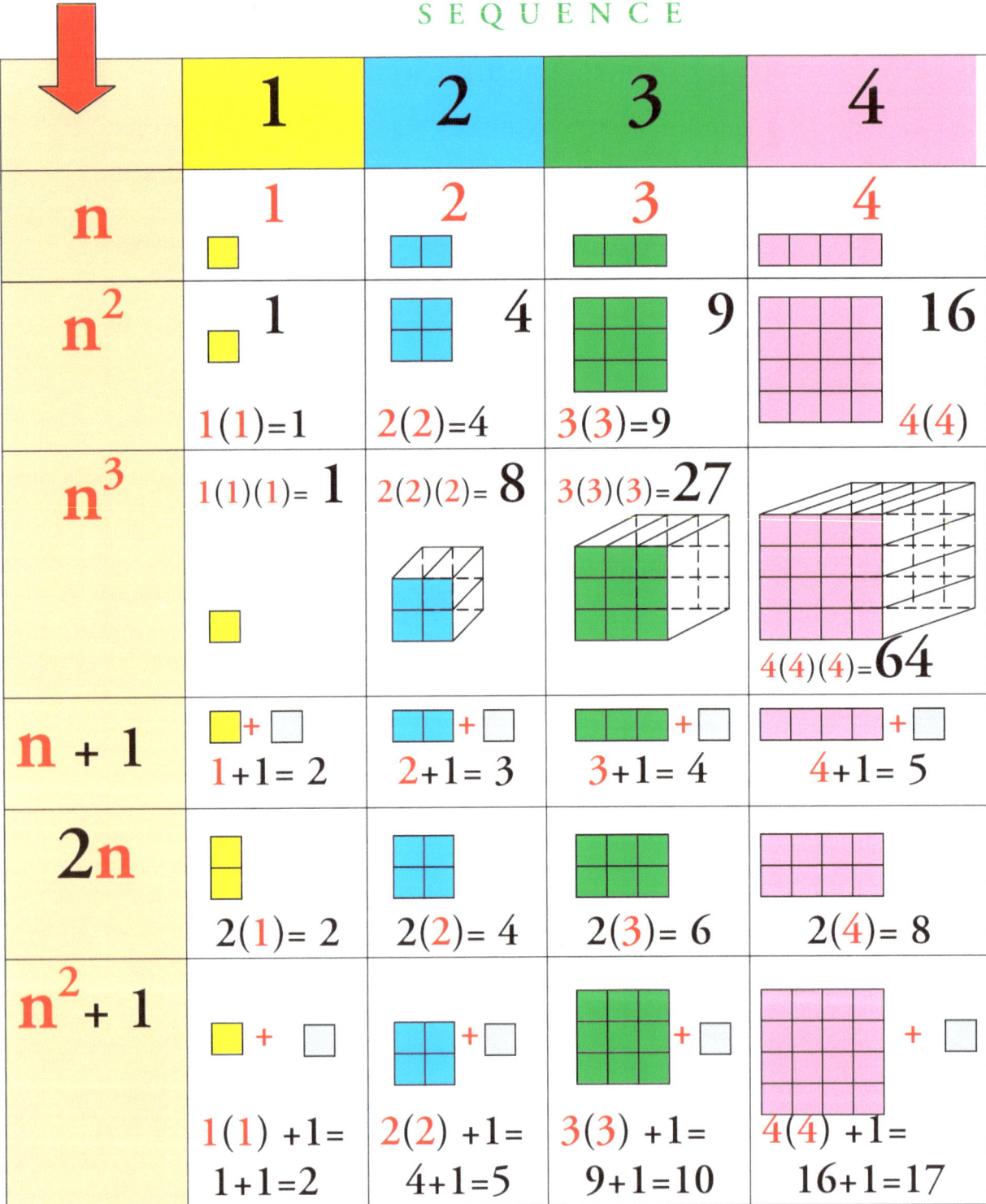

www.ingramcontent.com/pod-product-compliance
Lightning Source LLC
Chambersburg PA
CBHW051023180526
45172CB00002B/448